Second Edition
Thomas Gray & James Haynes assert their moral right to be identified as the authors of this work. The right of Thomas Gray & James Haynes to be identified as authors of this work has been asserted by them in accordance with the Copyright, Design and Patents Act 1988.

Illustrations by Thomas Gray
Page Makeup by Thomas Gray & James Haynes
Examples produced and checked by Thomas Gray & James Haynes
Foreword by Christopher Jeffery, Head, Bootham School, York

ISBN-13: 978-1976419799
ISBN-10: 1976419794

Sources and Images

Cover photograph: Skeeze (https://pixabay.com/)
Page 21: © James Haynes, George Johnson, Mike Shaw (Bootham School)

Images and Videos on Pixabay are released under Creative Commons CC0. To the extent possible under law, uploaders of Pixabay have waived their copyright and related or neighboring rights to these Images and Videos. You are free to adapt and use them for commercial purposes without attributing the original author or source. Although not required, a link back to Pixabay is appreciated.

- Frank Ward, First published in 2005 (latest edition published in Autumn 2008), *GCSE Astronomy Online,* http://www.gcseastronomy.co.uk, 14/10/16
- Baker J, Published 6th 2011, *50 ideas you really need to know: Universe,* London, Quercus, pp. 1-208.
- Ridpath I & Tirion W, New edition published 3 Mar. 2011, *Collins Stars & Planets,* Honley, Collins, pp. 1-400

Please direct inquiries to :

tom.gray@boothamstudent.co.uk
james.haynes@boothamstudent.co.uk

We would like to thank the Astronomy Department at Bootham School, firstly for their teaching of the course and then for their support during the production of this book .

Content

Specification: Understanding the Universe

Foreword

Chris Jeffery
Head, Bootham School, York

Three of the most notable features of Bootham School combined on my very first day here as Head.

Firstly, I got an introduction to the importance to the school of its long tradition of Astronomy. Blessed with its own observatory, an 1851 Thomas Cooke refractor telescope, and generous supporters among former students, as well, most notably, as staff of the skill, commitment, knowledge and enthusiasm of Mike Shaw and Russell Newlands. We have offered Astronomy at public examination level as one of our enrichment activities for many years.

Secondly, I encountered, for the first time, the remarkable confidence and generosity of spirit of Bootham students. Two of them, Tom and James, had just taken their GCSE's. At the tender age of 16, they had discovered that the commercially available revision material they had been working from had not been as helpful or as clear as it might have been. Rather than moaning about it, they decided to be the solution to this perceived problem... and turned up at my door with their own, improved book. Here it is!

Thirdly, they were offering their work not for commercial gain, but rather to the benefit of students from local schools who are themselves studying for a GCSE in Astronomy as part of the York Independent State School Partnership (a scheme which has proved a model nationally for such cross-sector collaboration). James and Tom are helping to tutor the current group, and are keen to ensure that the enthusiasts following in their footsteps can achieve their very best.

So, I commend this work to you, and extend my warmest and most grateful thanks to Tom and to James, to those who have so inspired them at Bootham and to those such as Annette Aylett whose passion for the excellent work of York ISSP ensures that this work will be put to good use.

I hope it proves useful and inspirational to generations of young astronomers in York and beyond.

Membra Sumus Corporis Magni
"We are members of a greater body"

Topic 1

Earth, Moon and Sun

Topic 1 Overview

Topic 1 is an introduction to Astronomy, relating astronomical and physical concepts to nearby astronomical bodies such as the Sun and Moon as well as the Earth itself.

Topic 1.1 - Earth: An in-depth study of the nature and origin of our Earth, in addition to some of the astronomical activity relating to it.

Topic 1.2 - Moon: An examination of the key observable features of our Moon. as well as the past and present methods of research used to collect data on the Earth's only natural satellite.

Topic 1.3 - Sun: An overview of the features both on and around the Sun as well as a variety of methods for observation.

Topic 1.4 - Solar System Interactions: An explanation of the key astronomical mechanics in play during phenomena such as aurorae, lunar phases and eclipses.

Key Facts and Figures

Earth
Diameter: 13,000km
Circumference of Earth: 40,000km
Mass: 5.972×10^{24} Kg
Rotational period: 23 hours 56 minutes
Orbital period: 365.256 days

Sun
Diameter: 1,400,000km
Distance from Earth: 150,000,000km
Length of sunspot cycle: 11 years
Rotational period: 25 days at equator, 31 days at the poles
Mass: 1.989×10^{30} kg

Moon
Diameter: 3,500km
Distance from Earth: 380,000km
Rotational period: 27.5 days
Orbital period: 27.5 days
Length of phase cycle: 29.5 days
Mass: 7.35×10^{22}

1.1

Earth

Planet Earth

Specification:

- Show an understanding of the Earth's water surface and atmosphere which differentiate it from other planets

- Explain how the Earth's sky is coloured blue by preferential scattering of light

- Display an understanding of the advantages of having an atmosphere

Unlike many planets, both in and outside of our Solar System, the Earth has liquid water on its surface which is thought to be fundamental to supporting life.

The Earth's rotational speed is a significant factor in the development of life. If it were slower, surface temperatures would significantly fluctuate depending on which side of the Earth the Sun's radiation hit. This would result in very hot and very cold areas.

The existence of an atmosphere around Earth has had a huge impact on the development of the planet and the organisms living on it. As well as providing the oxygen necessary for respiration, the atmosphere protects the Earth from collisions with astronomical bodies such as meteors and absorbs biologically harmful cosmic radiation.

The sky appears blue because white light from the Sun separates (scatters) into its different colours when entering the Earth's atmosphere. The shorter wavelengths (blue colours) pass through the gas particles whilst longer wavelengths (red colours) 'bounce off' oxygen and nitrogen molecules, making the sky appear blue.

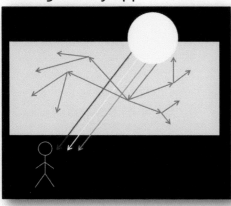

A diagram to show the preferential scattering of light as it enters the Earth's atmosphere

Shape

Specification:

- Know that the Earth's shape is roughly spherical

The Earth's shape is an oblate spheroid. This means a sphere which is slightly squashed by 42 kilometres at the poles. Gravitational forces act slightly more strongly on bodies at the poles than they do at the Equator.

The Earth is now known to be a sphere due to several inferred and direct reasons:

- Ships could only disappear over the horizon of a spherical Earth

- Satellites would only be able to orbit a spherical Earth

- Images from space missions have photographed and scanned a spherical Earth, proving its shape conclusively

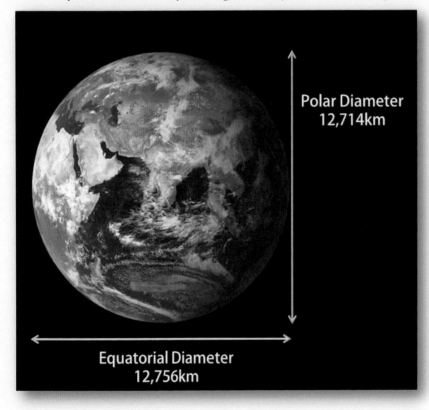

A diagram showing the Earth's polar diameter and equatorial diameter

Light Pollution

Specification:

- Describe what causes light pollution and explain why it is a problem for astronomers observing the night sky

Light pollution is defined as excessive amounts of natural and artificial light which impede observations of the night sky. It is caused by a variety of sources including street lights, house lighting, car headlights and sunlight reflected off the Moon.

Excessive amounts of light pollution can affect the eye's adaptation to the dark, which means that observers cannot see fainter stars with the naked eye. Light pollution can also cause a phenomenon called star glow, which makes the night sky appear an orange colour.

For this reason, visual observations in cities are difficult because of the higher density of light pollution Area of low population density favour astronomers looking at the night sky.

An example of light pollution and star glow caused by a house in the countryside

Atmosphere

Specification:

- Describe the composition of the Earth's atmosphere

- Compare the advantages and disadvantages of having an atmosphere

The Earth's atmosphere consists of a variety of gases. Nitrogen is the most abundant and makes up 78% of the overall volume. Additionally, the atmosphere comprises Oxygen (21%), Argon (1%), Carbon dioxide (0.04%) and traces of other noble gases (0.06%). The amount of water vapour present at any given time varies, therefore an exact value cannot be stated.

An understanding of both the benefits and potential drawbacks of our atmosphere from an astronomical point of view enables observations to be planned and executed successfully.

The Earth's atmosphere acts as a protective layer which helps to keep conditions on Earth conducive to life by absorbing harmful electromagnetic radiation, trapping oxygen, allowing the formation of liquid water and having a beneficial effect on surface temperature.

The atmosphere can often be a hindrance to astronomers. As waves of radiation penetrate the atmosphere many are absorbed, scattered and reflected. Preferential scattering gives the sky a blue appearance during the day, meaning that visual observations of the astronomical objects cannot be made continuously.

The Earth's atmosphere contains molecules such as water vapour which absorb electromagnetic waves at varying wavelengths, meaning that telescopes have to either be located on hilltops or on satellites so that the electromagnetic radiation coming from astronomical objects can be measured and observed with minimal interference from the atmosphere.

Eratosthenes

Specification:

- Explain how Eratosthenes was able to accurately calculate the circumference of the Earth

In about 240 BC, the Greek astronomer and mathematician Eratosthenes calculated the circumference of the Earth within a margin of error of 17%, using only his knowledge of the angle of elevation of the Sun.

He noted that on the 21 June, objects in Syene had no shadow at noon because the Sun was directly overhead whereas in Alexandria, 805 km away, objects did cast shadows. Eratosthenes then used the Sun's angle of elevation in Alexandria to calculate the circumference of the Earth as follows:

If ... 7°12' = 1/50 of a circle = 805 km
Then ... 50 x 805 km = 40,250km

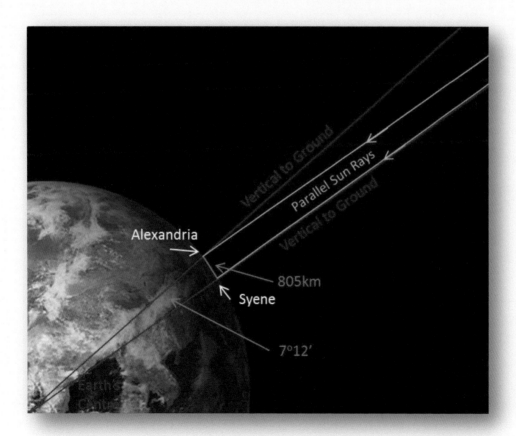

Terms

Specification:

- Define the terms: Equator, tropics, latitude, longitude, pole, horizon, meridian and zenith in an astronomical context

The following terms are all used when describing astronomical locations, phenomena or experiments:

Tropics – 23.5° north and south of the Equator. These are the two extreme latitudes that meet the ecliptic.

Poles – The points through which the axis of the Earth's rotation passes

Zenith – The point in the sky directly above an observer

Nadir – The point directly below an observer in relation to the celestial sphere

Horizon – The line where the land and sky touch

Meridian – A circle of set longitude that passes through both poles and the zenith of an observer. The prime meridian goes through Greenwich.

Latitude - The angle created between a point on the Earth's surface, the centre of the Earth and the Equator

Longitude – The angle created between a point on the Earth's surface, the meridian and the centre of the Earth

Equator – A line around the Earth which is equidistant from both poles. The Equator signifies where the Northern and Southern hemispheres touch.

Ecliptic – The apparent path that the Sun takes on the celestial sphere when observed from Earth.

Atmosphere and Waves

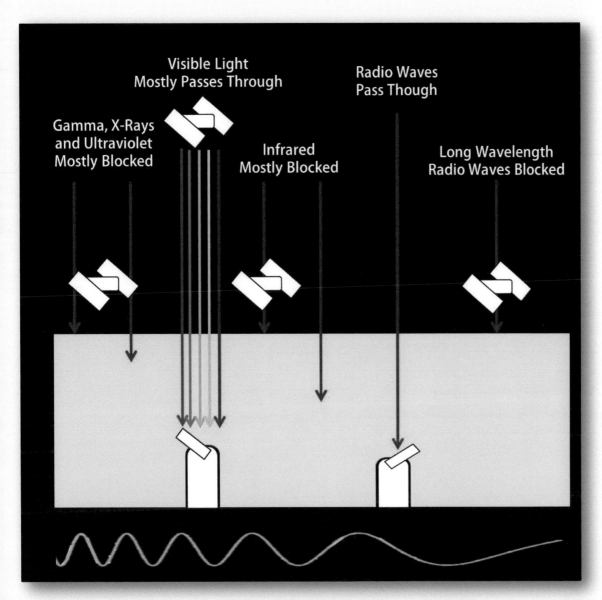

A diagram showing the varying penetrative power of waves in the electromagnetic spectrum as they collide with the Earth's atmosphere

Telescopes

Specification:

- Describe the main components of reflector and refractor telescopes

- Explain why the the largest telescopes are reflectors and not refractors

- Understand that visible light, microwaves and some radio waves are able to pass through the Earth's atmosphere

There are two types of telescope, reflectors and refractors. In both cases, the larger the telescope, the greater amount of light collected hence producing a higher resolution image.

Reflector: This telescope collects light through an opening and has a series of mirrors on which an image is reflected until it reaches the eyepiece. Larger telescopes are normally reflectors as large mirrors are easier to manufacture than the large lenses which are required in refractor telescopes.

Refractor: This telescope collects light which hits an objective lens and focuses an image onto a focal point which can be seen through an eyepiece. The distance between the focal point and the eyepiece is the focal length. The longer the focal length, the larger the image, so the casing is usually long and thin.

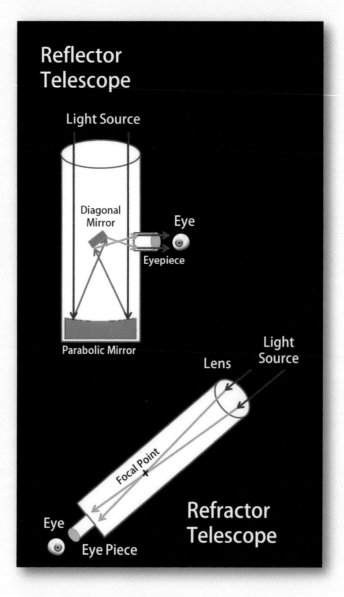

Spaced-Based Telescopes

Specification:

- Understand why most infrared, ultraviolet and x-ray observatories are sited on hills and are not located at Sea Level on Earth

- Compare the advantages and disadvantages of space-based telescopes

In general, shorter wavelengths such as gamma and x-rays, cannot penetrate the Earth's atmosphere. These spectrums can provide interesting data to astronomers, so several space-based telescopes have been launched beyond our atmosphere to take readings not possible from Earth.

Advantages of space based telescopes:

- There is no atmosphere to distort the clarity of visual images.
- There is no light pollution.
- Weather cannot interfere with observations.
- There are longer periods of darkness when observations can be made.

Disadvantages of space based telescopes:

- The lifetimes of space based telescopes are relatively short.
- It is almost impossible to upgrade and maintain telescopes on satellites.
- There is an additional cost to launch the telescope.
- There is an additional environmental impact.

Hubble Space Telescope (left) , Nustar Space Telescope (right)

Van Allen Belts

Specification:

- Describe the discovery of the Van Allen Belts and their composition

The **Van Allen belts** are two bull-horn shaped sets of rings of high-energy particles which are held in place by the Earth's magnetic field. They were formed due to the interaction of cosmic waves with the Earth's atmosphere.

The Inner Belt

The inner belt was discovered in 1957 by the Explorer 1 probe which detected high-energy particles with a Geiger counter. The discovery was confirmed by Explorer 3 and, later, Sputnik 3.

The inner belt is mostly made up of high-energy protons at low altitude between 600 and 1000 km above the Earth.

The Outer Belt

The outer belt was discovered in 1958 by Pioneer 3. The probe failed at 100,000 km above the Earth, but managed to detect the outer belt before burning up. The discovery was confirmed a year later by Pioneer 4.

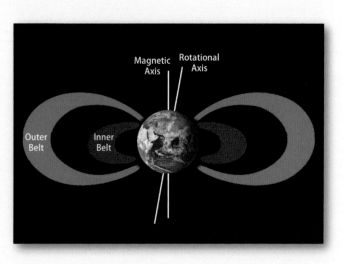

The outer belt is mostly made up of high energy electrons and is less dense than the inner belt. The electrons and other charged particles are thought to originate from solar wind thrown out by the Sun. The outer belt has wider 'horns' that dip in at a much steeper angle than the inner belt and is much higher, reaching between 15,000 and 65,000 km above the Earth.

1.2

Moon

Figures

Specification:

- Recall the key figures including the Moon's diameter (3,500km) and its distance from Earth (380 000 km)

Our Moon is the Earth's only natural satellite. It is on average 380,000 kilometres from Earth which is about 400 times smaller than the distance between the Sun and Earth. The Moon's diameter is about 3,500 kilometres which is around 1/400 the size of the Sun's diameter. These two ratios allow for the occurrence of solar eclipses which are thought to be relatively rare in the Universe.

The Moon is also responsible for natural phenomena such as changing tides, waves and a slightly longer length of day.

Astronomers have always been interested in studying the Moon with ancient Greek/Roman scientists hypothesizing that the dark patches visible on the Moon's surface were oceans.

Image showing a Full Moon. Features like craters and rilles can be clearly seen

Lunar Map

Specification:

- Be able to label some of the Moon's main features, for example the Sea of Tranquility, Ocean of Storms, Sea of Crises, the craters Tycho, Copernicus and Kepler and the Apennine mountain range

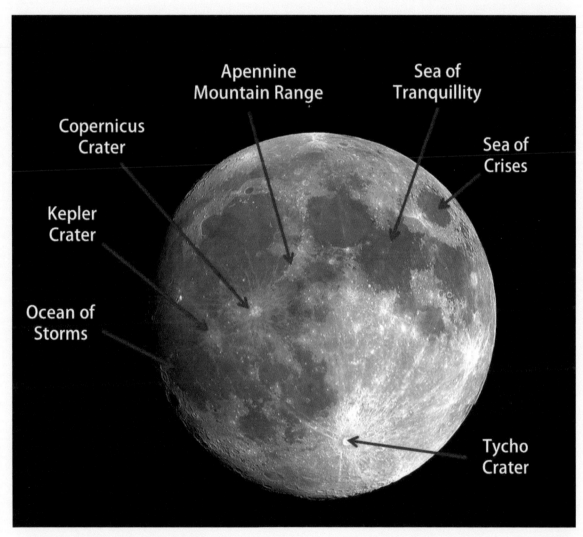

Image showing a Full Moon

Features

Specification:

- Describe the nature, composition and origin of rilles and wrinkle ridges

Common features on the Moon's surface are rilles which appear as long, snaking groove-like structures which can run for several kilometres. Rilles are thought to have formed from drained lava tunnels whose roofs collapsed.

Another common feature are wrinkle ridges which are the equivalent of the Earth's hills. They can be several hundred meters high and thousands of kilometres long. Wrinkle ridges formed as the the Moon's outer crust cooled and expanded.

(Top) Photographs of the near side lunar surface clearly showing wrinkle ridges, rilles and craters. Technical information: Photograph produced using a webcam attached to telescope

(Bottom) During the Apollo 15 mission astronauts used a lunar rover to explore the surface of the Moon, most notably the Hadley Rille. While surveying the rille astronauts nearly collided with it

Craters

Specification:

- Explain the origin of the craters on the Moon's surface

The Moon has no atmosphere so even the smallest asteroids are able to impact its surface, leading to its heavily cratered appearance which is still observable today. The majority of these craters are thought to have been made during a period known as the 'heavy bombardment' by objects left in the Sun's accretion disk.

Craters are best observed when the Moon is not full because when light hits the crater at an angle, the raised rim casts a shadow which emphasizes the centre's depth and makes calculation of the diameter and depth of the crater easier with a smaller margin of error.

Bright streaks are visible around large craters which are most likely to be material underneath the Moon's darker surface which was thrown out on impact.

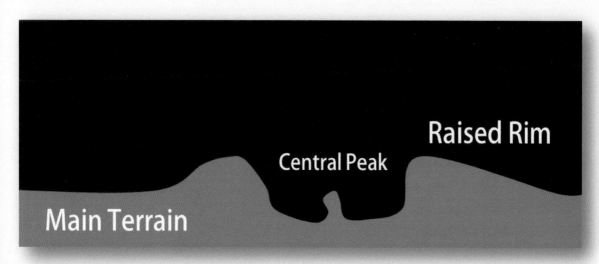

A diagram to show the cross-section of a crater on the Moon

Maria & Terrae

Specification:

- Explain the origin and nature of the lunar seas (maria) and lunar highlands (terrae)

- Explain how the relative ages of lunar features can be determined by the number of craters on maria and terrae

Maria

The period of heavy bombardment of the Moon was followed by a period of volcanic activity, when lava leaked into the newly formed craters and filled them. As the lava cooled and solidified, large flat, dark plains called maria were formed. Around 17% of the Moon's surface is covered in maria. There are fewer maria than terrae on the far side of the Moon.

Terrae

Terrae are the highland areas of the Moon. Terrae are much higher and appear much lighter than maria. Many more craters are visible on terrae than on maria, which suggests that terrae are significantly older.

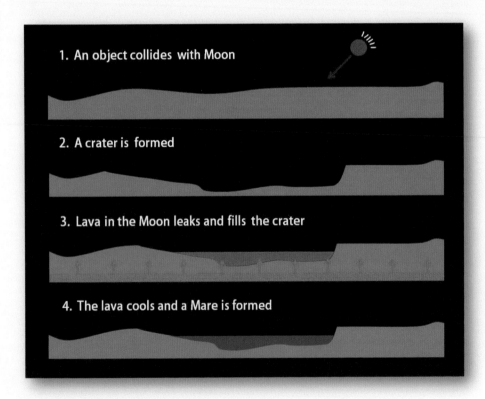

1. An object collides with Moon

2. A crater is formed

3. Lava in the Moon leaks and fills the crater

4. The lava cools and a Mare is formed

Diagram explaining the formation of a mare on the lunar surface

Orbit & Rotation

Specification:

- Explain why the Moon's rotational period and orbital period are both 27.3 days

The Moon has a rotational period of 27.3 which is the same as its orbital period due to the gravitational tidal effect from Earth. This results in the Moon having a synchronous orbit with the Earth, meaning that the 'far side' of the Moon is never visible. The Moon used to rotate faster, but has gradually slowed down.

The Moon takes 29 days rather than 27.3 days to return to the same position in the sky relative to observation from Earth due to the movement of the Earth around the Sun at the same time that the Moon is orbiting around the Earth - it takes the extra 1.7 days for the Moon to 'catch up'.

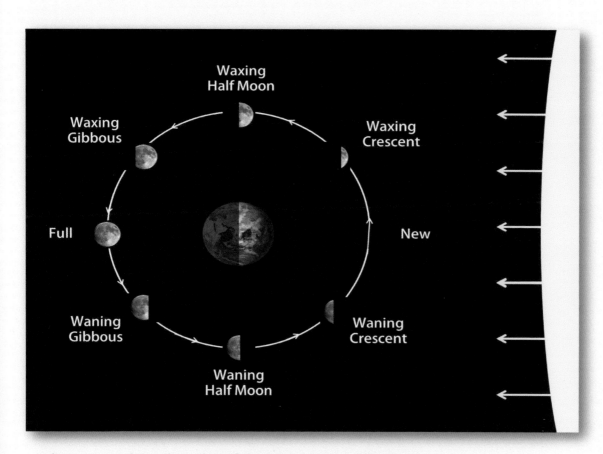

A diagram to show the orbit of the Moon around the Earth

Far Side of the Moon

Specification:

- Explain why the far side of the Moon is not observable from Earth

- Explain how astronomers were able to discover the nature of the far side of the Moon with modern technology

The Moon has a rotational period and orbital period that are the same. As a consequence, only one side is ever visible from Earth, meaning that until relatively recently astronomers knew nothing of the far side's appearance.

The far side was first photographed by the United Soviet Socialist Republic (USSR) probe Luna 3 in 1959. This allowed astronomers and geologists to obtain information about its appearance and structure. Since then many other probes have taken photographs allowing further, more detailed observations to be made.

Clear differences can be observed between the far side and the near side of the Moon. The far side has notably fewer lunar seas due to a thicker surface crust which has stopped molten rock piercing through. The far side surface has many more craters due to a greater number of impacts from space. The near side is protected from impacts by the position of the Earth..

A photograph showing the visible side of the Moon

Gravity on the Moon

Specification:

- Relate the lack of atmosphere to the Moon's low gravity

Gravity on the Earth is 9.8 N/kg whereas on the Moon it is only 1.6 N/kg. This difference means that on Earth gases are trapped and form an atmosphere, whereas on the Moon this cannot happen. Another consequence of the difference in gravitational field strengths is that objects weigh more on the Earth than on the Moon.

The Moon's lack of atmosphere makes it a perfect site for telescopes which can observe space in wavelengths which would be absorbed by the Earth's atmosphere. The Moon's remoteness also reduces interference from signal-creating devices such as mobile phones.

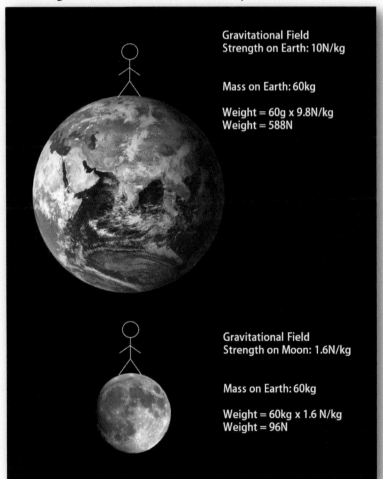

Gravitational Field
Strength on Earth: 10N/kg

Mass on Earth: 60kg

Weight = 60g x 9.8N/kg
Weight = 588N

Gravitational Field
Strength on Moon: 1.6N/kg

Mass on Earth: 60kg

Weight = 60kg x 1.6 N/kg
Weight = 96N

Diagram illustrating the differences between gravity on the Moon and the Earth

Moon Exploration Missions

Specification:

- Explain the significance of the Apollo space programme

- Explain how ALSEPs were used to carry out hugely significant experiments to gain a better understanding of the Moon as well as our Universe

Between 1969 and 1972 six lunar landings were carried out by the North American Space Agency (NASA) which were named the Apollo missions and given numerical designations. The main goal of the Apollo missions was to land men safely on the Moon and return them to Earth. The first successful manned landing was Apollo 11, with the previous 10 missions testing American space capability.

As well as achieving the first Moon landing, the Apollo missions conducted a number of scientific experiments using the Apollo Lunar Surface Experiments Package (ALSEP) which comprised a series of geophysical instruments which ran autonomously and gathered data about the environment of the Apollo landing sites after the astronauts had left. This data stream, along with Moon rock samples, were then used by NASA to better understand the Moon as it is currently and allowed them to hypothesise about the Moon's origins.

The ALSEPs included instruments to:

- Measure seismic activity on the Moon
- Detect charged particles in solar wind
- Detect a very thin lunar atmosphere
- Detect a laser beam fired from Earth to measure the distance accurately
- Measure the Moon's atmosphere accurately

An astronaut taking a sample of lunar rock for an ALSEP experiment

Origins of the Moon

Specification:

- Explain how the Moon was probably formed through the giant impact hypothesis

- Explain how evidence was used to validate the giant impact hypothesis

The exact origins of the Moon are unknown, however there are many plausible theories. The most widely accepted theory is known as the giant impact hypothesis, which suggests that the Moon was created over four billion years ago during a collision between Earth and a Mars-sized object. As the two bodies collided, they fused whilst simultaneously ejecting debris from the opposite side which combined to form the Moon. The Moon acts as a gravitational counterweight to the Earth and allows it to maintain its steady rotational period.

Evidence has been acquired which supports this hypothesis. During the 1969-72 Apollo Missions, rocks were retrieved from the surface of the Moon which on analysis were found to contain the same relative proportions of certain oxygen isotopes as found on Earth.

The Moon's surface is also known to have been molten at one time, which would have required a large amount of energy. This could have originated from a sizeable impact.

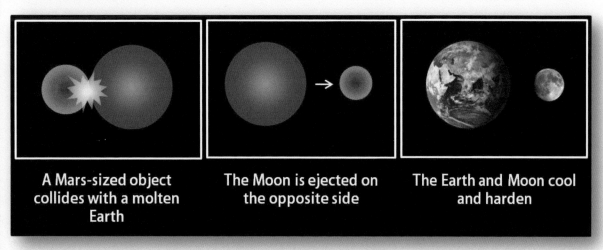

| A Mars-sized object collides with a molten Earth | The Moon is ejected on the opposite side | The Earth and Moon cool and harden |

1.3

Sun

Rotation
Observation
Wavelengths
Layers
Sunspots
Fusion

Solar Rotations

Specification:

- Recall that the Sun's rotation period varies from 25 days at the equator to 36 days at its poles

- Demonstrate an understanding of how astronomers use observations of sunspots to determine the Sun's rotation period

Our Sun is the star that lies at the centre of our Solar System. Its average diameter is 1,400,000 km and it is located 150,000,000 km (one Astronomical Unit) away from Earth.

The Sun's body consists of a complex combination of plasma and gas. This leads to a varying rotation. At the equator the rotational period is 25 days while at the poles it can be anything up to 36 days. This variation occurs due to the Sun's massive size and flexible structure. The Sun's rotational period can be calculated by simply recording the relative speed and position of various sunspots as they traverse the photosphere.

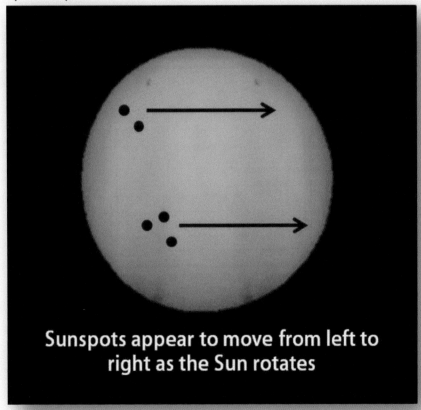

Sunspots appear to move from left to right as the Sun rotates

Observations

Specification:

- Describe how the Sun can be safely observed by amateur astronomers

To prevent damage to the eye when observing the Sun, either protective filters must be used on binoculars and telescopes or an indirect method of observation must be set up.

The Sun can be safely observed with a telescope fitted with a solar filter such as a hydrogen alpha filter.

Alternatively, a number of indirect observation techniques can be used such as viewing the image of the Sun projected onto a canvas. This involves pointing a telescope at the Sun and observing the projection of the Sun on a screen. This method is safe and allows sunspots and eclipses to be easily observed.

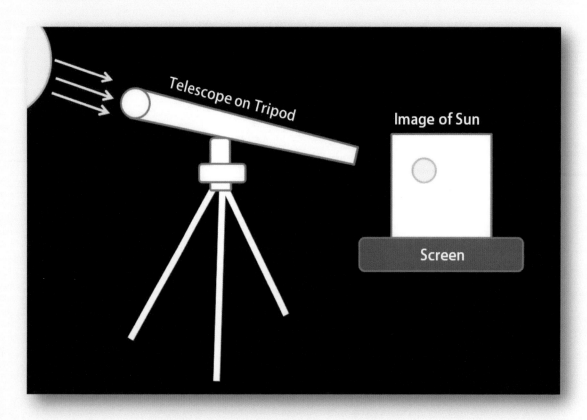

Wavelengths

Specification:

- Demonstrate an understanding of the appearance of the Sun at different wavelengths of the electromagnetic spectrum, including visible, H-alpha and x-ray

- Describe how astronomers observe the Sun at different wavelengths

The Sun's core reaches extremely high temperatures which releases electromagnetic waves across the spectrum. These different frequencies give the Sun a completely different appearance depending on how it is being observed.

Observations in different spectrums have particular advantages and can reveal and highlight certain features of the Sun, for example solar flares and sunspots are difficult to view in detail using just visible light.

The Hydrogen-Alpha spectrum is frequently used by astronomers when observing the Sun as it allows a good view of the surface as well as the solar prominences.

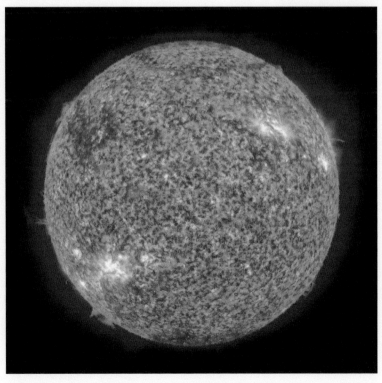

Layers

Specification:

- Describe the order of the different layers of the Sun from core to corona

- Recall the changing temperatures in the Sun from layer to layer

- Describe the location from which key solar features originate including solar prominences, sunspots and flares

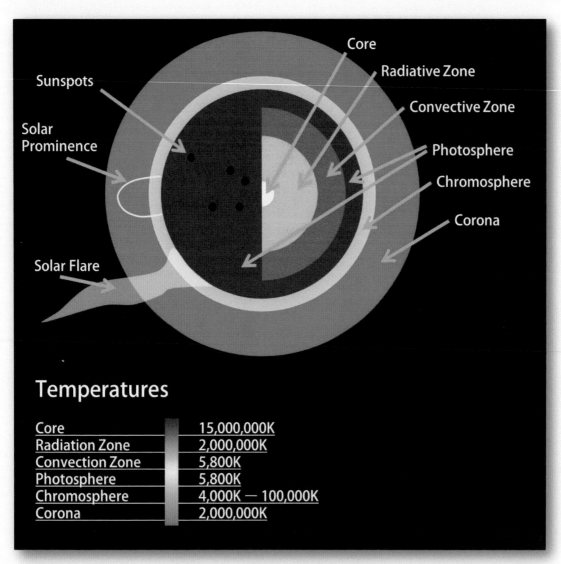

Temperatures

Core	15,000,000K
Radiation Zone	2,000,000K
Convection Zone	5,800K
Photosphere	5,800K
Chromosphere	4,000K — 100,000K
Corona	2,000,000K

A diagram to show the layers and features of the Sun

Sunspots

Specification:

- *Describe the nature and appearance of sunspots*

- *Recognise the shape of a butterfly diagram and describe the latitudinal drift of sunspots towards the Sun's equator*

- *Use butterfly diagrams to predict the next period of solar maximum*

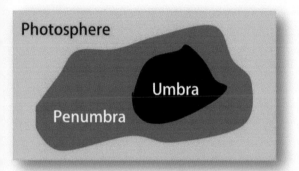

Sunspots are cooler areas on the photosphere of the Sun which gives them a dark appearance. They are normally around 50,000 km wide and occur between 40° north and 40° south. They occur due to changes in the Sun's magnetic field.

Sunspots appear as a very dark central umbra (4000K), surrounded by a lighter penumbra (5600K). A stream of high-energy particles in the form of solar wind is released when the Sun is producing large amounts of heat.

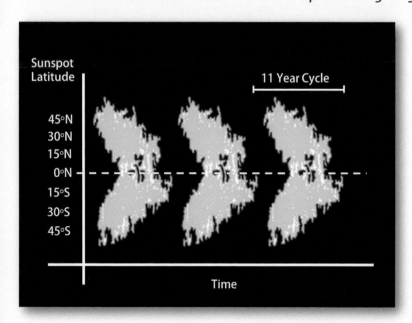

The distribution of sunspots is predictable, because they move periodically towards the equator of the Sun in an eleven- year cycle which can be mapped on a 'butterfly diagram', shown to the left.

Fusion

Specification:

- Explain how the Sun generates energy through nuclear fusion

In the extreme temperatures reached inside the Sun's core, hydrogen nuclei are able to fuse into larger helium nuclei in a series of nuclear fusion reactions called the proton-proton chain. This simultaneously releases large amounts of energy which travel towards the surface of the Sun through convection currents.

When all the hydrogen nuclei have been used up, the proton chains continue by fusing helium nuclei until the iron stage is reached (elements heavier than iron are formed in supernovae).

The complete cycle of helium nuclei fusion can be seen in the diagram below:

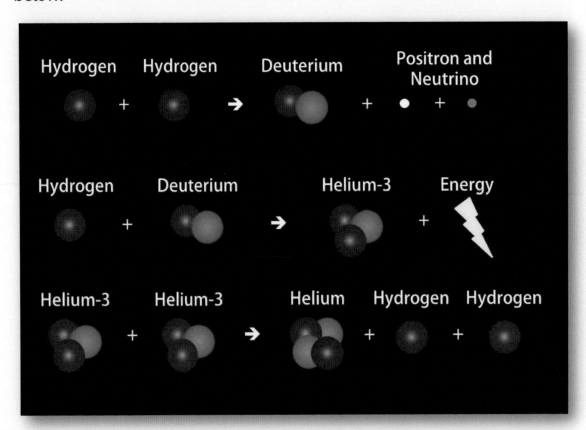

1.4

Interactions

Sizes

Specification:

- Explain why it is possible to observe the Moon eclipsing the Sun due to their relative sizes and position in the Solar System

The Moon is roughly 380,000 km from Earth and has a diameter of around 3500 km. The Sun is 150,000,000 km from Earth and has a diameter of around 1,400,000 km.

This means that the ratio between the diameter of the Moon and the diameter of the Sun is around 1:400, which is the same ratio as their distances to the Earth as is demonstrated in the diagram below. This coincidence allows for phenomenon of a solar eclipse, as the two bodies appear to be the same size when observed from Earth.

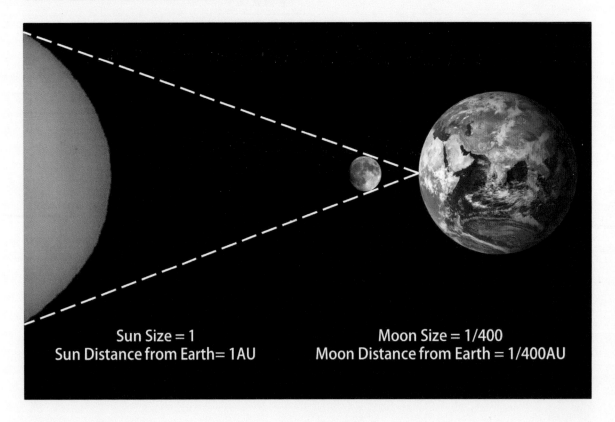

Sun Size = 1
Sun Distance from Earth = 1AU

Moon Size = 1/400
Moon Distance from Earth = 1/400AU

Lunar Phases

Specification:

- Know that the Moon's phase cycle is 29.5 days long

- Be able to identify what phase the Moon is in based on what it looks like from an observatory on Earth

- Explain, with a diagram if necessary, why the Moon's phase cycle is 2.2 days longer than its orbital period

When observed from the Earth, the Moon appears to go through different phases due to different parts being hit by sunlight during its orbit around the Earth.

The Moon's phase cycle is 29.5 days, 2.2 longer than its orbital period. This is because the Earth is orbiting around the Sun while the Moon is orbiting around the Earth and it therefore takes extra time for the Moon to catch up with the Earth and reach the same position as in the last cycle.

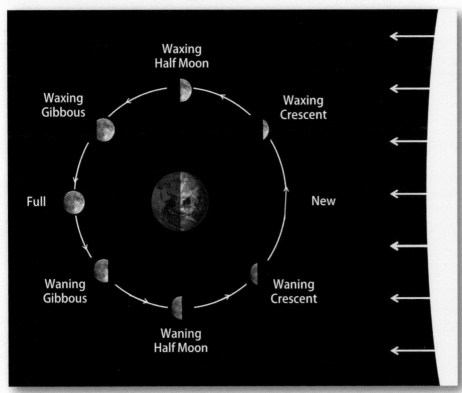

A diagram to show the orbit of the Moon around the Earth

Lunar Eclipses

Specification:

- Describe, with a picture if necessary, the appearance of the Moon during a lunar eclipse

- Explain, with a diagram if necessary, how a lunar eclipse forms

- Explain why a lunar does not occur every Full Moon

A lunar eclipse can be observed when the shadow of the Earth falls onto the Moon. The Earth's shadow is much larger than the Moon's, so lunar eclipses happen more frequently than solar eclipses.

As lunar eclipses can only happen when the Earth is between the Moon and the Sun, the Moon will always appear full during the eclipse. During a total lunar eclipse, the Moon appears a blood red colour due to Earth's Atmosphere refracting sunlight. Eventually the Moon disappears completely into shadow and then reappears with the same red colour before returning to normality. The Moon does not orbit on the same plane as the Earth so there is not a lunar eclipse every full moon.

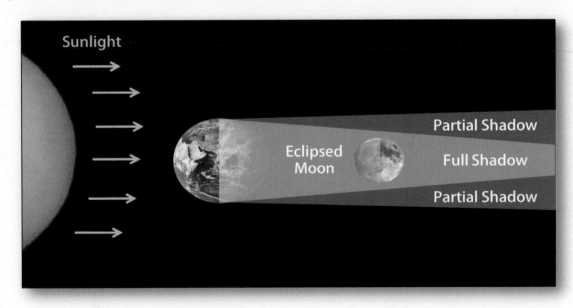

Solar Eclipses

Specification:

- *Describe, with a picture if necessary, the appearance of the Moon during a solar eclipse*

- *Explain, with a diagram if necessary, how a solar eclipse forms*

- *Explain why certain parts of an area in shade only experience a partial eclipse and why a solar eclipse does not occur every Full Moon*

A solar eclipse can be observed when the Moon passes between the Sun and the Earth and the Moon partially or fully occults the Sun. This happens quite rarely as the plane on which the Moon revolves around the Earth is not on the ecliptic so the two bodies do not cross in the sky once a month as may be assumed. During a total eclipse, the Moon gradually covers the Sun. When the coverage is nearly complete, a 'diamond ring effect' is visible (see right) due to the presence of topographical features on the Moon's surface which allow light to pass.

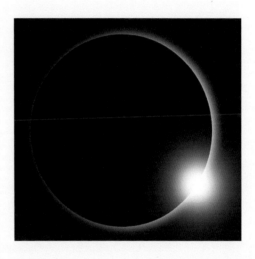

During totality, which lasts for about a minute, the Sun's corona can be seen as a glow around the ring. Then another diamond ring effect is visible on the opposite side of the Moon, followed by the receding of the Moon's shadow until the Sun returns to normality.

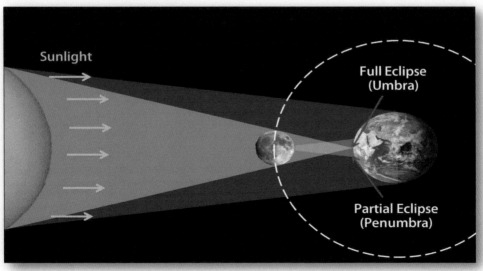

The Day

Specification:

- *Define the terms solar day and sidereal day*

- *Explain why a solar day is four minutes longer than a sidereal day*

- *Explain how a sundial is used to determine time*

Solar Day: The time between two consecutive transits of the meridian by the Sun from a given location

Sidereal Day: The time between two consecutive transits of the first point of Aries by the Sun. The sidereal day is 23 hours and 56 minutes.

The sidereal day represents how long the Earth takes to rotate 360° on its axis and so is relative to stars which do not appear to move from Earth. The solar day is therefore longer as it takes an additional four minutes to rotate more than 360° due to the Earth's orbit around the Sun.

A sundial can be used to determine actual time. The changing position of the Sun during the day causes a shadow with a changing position to fall on a series of numbers inscribed on a dial face. The style, which is the time-telling edge of the shadow-casting part of the sundial, must be parallel to the axis of the Earth's rotation for the sundial to be accurate throughout the year, meaning that the angle from the horizontal will always be that of the observer's latitude.

Time

Specification:

- Explain the difference between 'apparent solar time' and 'mean solar time'

- Explain how time zones are used to standardise time in regions

Time calculated with a sundial is called **apparent solar time** and represents time based on the actual position of the Sun, however due to the Earth's elliptical orbit and the tilt of its axis, solar time taken from a sundial can be inaccurate by up to 16 minutes.

To standardise time in set time zones, **mean solar time** was introduced to represent a model of the Sun if it moved across the sky at an even rate throughout the year, which allows for a standardised model of time.

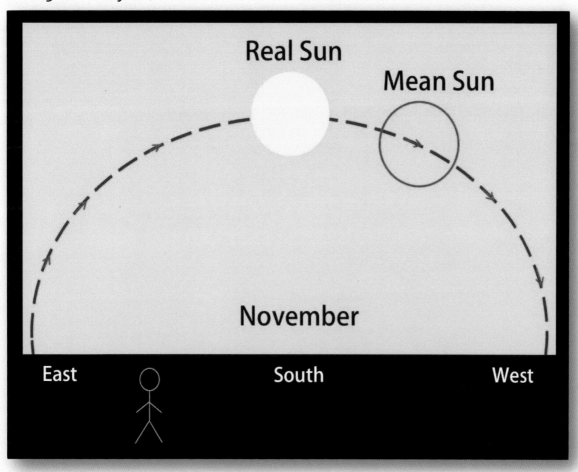

The Equation of Time

Specification:

- Understand that sunrise and sunset times vary with the time of year and that as a result, the number of daylight hours varies

- Understand the 'equation of time' (EoT) and its role in time standardisation

The equation of time (EoT) is the measured difference between apparent solar time and mean solar time. Apparent solar time is the time between two successive crossings of the meridian by the Sun. This can be measured using a sundial (or other similar instruments) as the time is displayed in the form of a shadow. Mean solar time is the average time throughout the year for the Sun to do this and is the time displayed on your watch (GMT).

During the year the equation of time value changes and at points can be up to 16 minutes either side of noon. This happens because of the Earth's changing 'tilt' as it revolves around the Sun, meaning that different hemispheres are angled by differing degrees towards the Sun. This angle changes over time and causes the changing seasons.

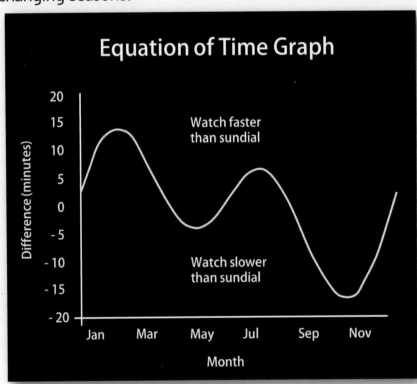

Aurorae

Specification:

- Describe the appearance of aurorae and where they are observable

- Explain the processes which create aurorae

Aurorae appear as bright red, blue, green or crimson streaks of light in the sky and can be observed in polar regions. The lights normally form at an altitude between 100 and 1000 km. Aurorae are caused by oxygen and nitrogen molecules in the upper atmosphere becoming excited by electrons released by solar wind which were accelerated to high speeds by the Earth's magnetic field. As the molecules de-excite, they emit light at different wavelengths depending on the element.

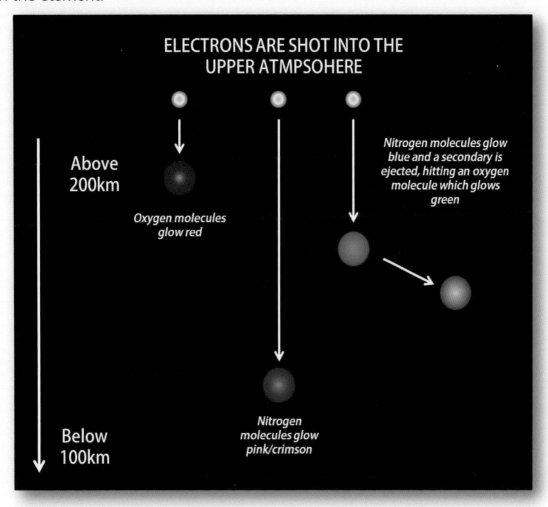

Topic 2

Planetary Systems

Topic 2 Overview

Topic 2 investigates the further reaches of the Solar System including the main planets and dwarf planets as well as features such as the Asteroid Belt, Kuiper Belt and Oort Cloud. The topic also covers details about smaller astronomical bodies such as asteroids, comets and meteors including their origin and nature. The second half of topic 2 covers the past and present advances made by astronomers when exploring our Solar System and then goes on to discuss the process of discovering other Solar Systems and their planets.

Topic 2.1 - The Solar System: An overview of the key terminology used by astronomers when describing the Solar System and the location of its features.

Topic 2.2 - Comets and Meteors - A detailed study of two of the most common astronomical bodies in our Solar System which includes a look at the potential danger the Earth faces from them.

Topic 2.3 - Solar System Discoveries: A history of the most important advancements made by contemporary astronomers such as Copernicus, Kepler and Galileo as well as case studies of famous astronomical discoveries.

Topic 2.4 - Exoplanets: An examination of the nature and discovery of exoplanets as well as a discussion about the possible advantages and disadvantages of contact with intelligent life outside our Solar System.

Key Facts and Figures

Planets in our Solar System from closest to furthest away from the Sun

Mercury, Venus, Earth, Mars, Jupiter, Saturn, Uranus, Neptune

Planets in order of size from smallest to largest

Mercury, Mars, Venus, Earth, Neptune, Uranus, Saturn, Jupiter

Dwarf planets in our Solar System from closest to furthest away from the Sun

Ceres, Pluto, Haumea, Makemake, Eris

Dwarf planets in order of size from smallest to largest

Eris, Pluto, Haumea, Makemake, Ceres

2.1

The Solar System

Terms

Specification:

- Define the terms greatest elongation, inferior planet, superior planet, conjunction, opposition, transil, occultation and greatest elongation

Inferior Planet is a planet that is closer to the Sun than the Earth, for example Venus and Mercury

Superior Planet is a planet that is further away from the Sun than Earth, for example Mars, Jupiter, Saturn, Uranus and Neptune

Inferior conjunction occurs when the Earth and another planet are in line with the Sun and are on the same side of the Sun.

Superior conjunction occurs when the Earth and another planet are in line with the Sun but are on opposite sides of the Sun.

Opposition occurs when the Earth is positioned between a planet and the Sun and is the best time to observe the planet because at this point it is closest to the Earth.

Transit occurs when one planet passes in front of another. Usually used to describe when Venus or Mercury pass in front of the Sun. Transits of Mercury are much more common than those of Venus.

Occultation occurs when a transit blocks out the light from another body. Normally used to describe the Moon obscuring a planet or a star. An eclipse is another common example of an occultation.

Greatest elongation occurs when an inferior planet forms a 90° angle between the Sun and Earth. The remaining angle that the Earth forms between the Sun and the inferior planet is known as the angle of elongation.

All of these terms have been described relative to Earth, but can be used to describe the position of any planet in any solar system

Terms

Specification:

- Label a diagram to illustrate the terms conjunction, greatest elongation and opposition

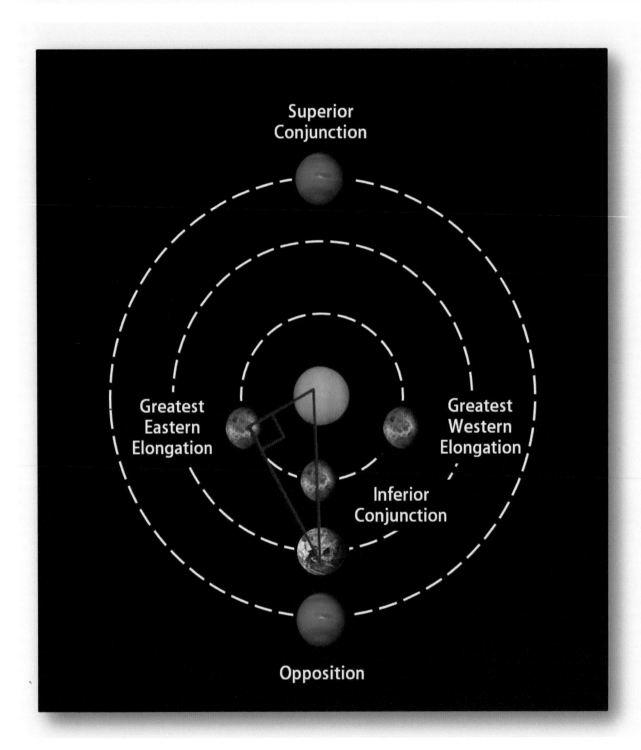

The Ecliptic

Specification:

- *Understand what is meant by the 'ecliptic'*

- *Understand that planets in our Solar System move in elliptical orbits, inclined to the ecliptic plane*

- *Understand that planets in our Solar System are observable in the zodiac band around the ecliptic*

All planets in our Solar System orbit on the same plane around the Sun called the ecliptic. Due to the Earth's axis being tilted at 23.5° from the ecliptic, when observing the Sun from the Earth, the Sun appears to travel along the ecliptic across the sky at 23.5° to the horizon.

From our perspective, the ecliptic runs through 13 constellations which were named by the ancient Greeks, 12 of which are the modern day astrological zodiac signs. 9° either side of the ecliptic is the zodiacal band where planets can be found when observed from the Sun. The angle at which planets deviate from 23.5° is their inclination.

The Earth's axis also 'wobbles' to some degree which is called precession.

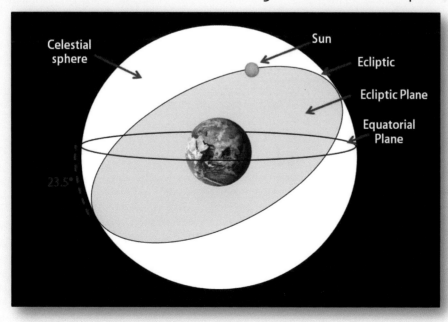

Orbits and Units

Specification:

- Be able to visualise our Solar System in terms the sizes of planets and the distances between them

Object	Diameter (km)	Distance (AU)	Distance (million km)
Sun	1,400,000	-	-
Mercury	4,900	0.4	57
Venus	12,100	0.7	108
Earth	13,000	1.0	150
Moon	3,500	-	-
Mars	6,800	1.5	228
Jupiter	142,800	5.2	778
Saturn	120,000	9.5	1,427
Uranus	51,000	19.2	2,871
Neptune	49,500	30	5,914
Kuiper Belt	-	100-1000	-
Oort Cloud	-	10,000 - 20,000	-

Orbits and Units

Specification:

- *Understand what is meant by an 'astronomical unit'*

- *Describe the nature of Earth's orbit around the Sun*

- *Describe the relative sizes of the Earth, Moon and Sun using the everyday items*

The Earth revolves around the Sun in an elliptical orbit (a very slight oval) which means that the distance between the Earth and the Sun is variable. The point in the Earth's orbit when the Earth is closest to the Sun occurs in January and is called aphelion. The point when the Earth is farthest away from the Sun occurs in July and is called perihelion.

For this reason, the mean orbit of the Earth was created to be a simplified model where the circumference is the same and the distance between the Earth and the Sun is a constant throughout the orbit. This distance is 1 Astronomical Unit (AU) and is around 150,000,000 km.

If shrunk down to a very small scale the Solar System can be visualised as a basketball (the Sun), a glass marble (the Earth) and sago (the Moon).

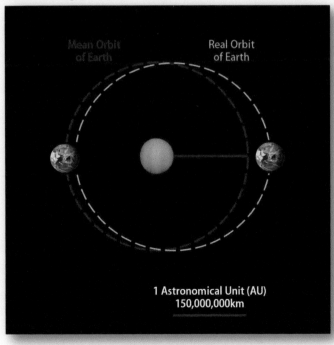

Motion

Specification:

- Demonstrate an understanding of the direct and retrograde motion of planets on a star chart

Normally when observing the movement, or planetary track, of a superior planet, the path will be from east to west in direct motion. However sometimes the planet will appear to remain stationary, or even move backwards, which is known as retrograde motion.

This occurs because the Earth is closer to the Sun than a superior planet, so it has a smaller orbit. The planet therefore may appear to be moving in direct motion, but as the Earth catches up and aligns, the planet will appear to be stationary for a while. Then as the Earth overtakes the planet it will appear to move backwards for some time in retrograde motion.

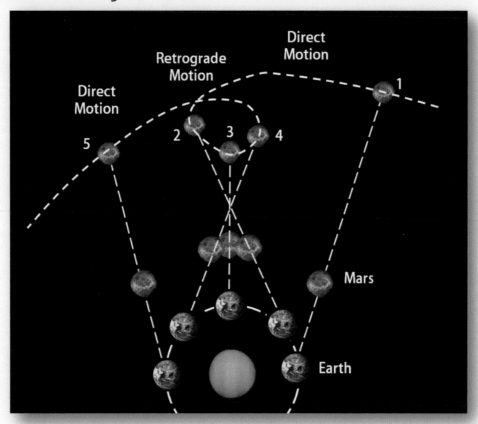

A diagram to show the stages of direct motion, stationary point and retrograde motion shown when Mars is observed from Earth

Planets

Specification:

- *Recall the names of planets in our Solar System and be able to list them in order of distance from the Sun*

- *Describe the key physical and chemical features of each planet including surface features, atmospheric composition and temperatures*

Planet	Surface Features	Atmosphere	Temperature	Notes
Mercury	Craters and cracks similar to the Moon	No atmosphere	440K facing Sun	Thick rock crust, large iron core creates small magnetic field
Venus	Geologically young, evidence of volcanic structures	Dense atmosphere - 96% Carbon Dioxide plus Sulphuric Acid	750K average	Not possible to view due to thick cloud cover
Earth	70% water, geologically active, supports life	Medium atmosphere - 78% Nitrogen 21% Oxygen	290K average	Crust movement through plate tectonics, inner and outer core create magnetic field
Mars	Active weather, but geologically extinct, thick crust and possible flowing water in past	Thin atmosphere 95% Carbon Dioxide 3% Nitrogen	220K average	Low density core, largest mountains and canyons in Solar System
Jupiter	Gas giant, no solid surface, bands of cloud visible, very thin ring system	90% Hydrogen 10% Helium	125K average at cloud top	Largest planet in Solar System, red spot is Earth sized storm, flattened sphere shape from fast rotation
Saturn	Gas giant, large ring system	96% Hydrogen 4% Helium	95K average at cloud top	Very low density
Uranus	Green gas giant, few features, dark ring system	83% Hydrogen 15% Helium 2% Methane	55K average at cloud top	Small ice-rock core, has an inclined axis so it appears to rotate on its side
Neptune	Blue gas giant, thin clouds, active weather system, very thin ring system	79% Hydrogen 18 % Helium 3% Methane	55K average at cloud top	Small ice-rock core, internal heat source

Trans-Neptunian Objects

Specification:

- Recall the nature and location of Trans-Neptunian Objects

Trans-Neptunian Objects are large astronomical bodies that orbit beyond the orbit of the planet Neptune. Pluto was the first such object to be discovered, however over 1000 TNOs have been discovered since then.

Some of the objects have 'resonant orbits' which means that their shape is similar to the orbit of Neptune. It is very difficult to observe TNOs as they are very distant and reflect very small amounts of light.

Pluto is one of the largest known trans-Neptunian objects. Others include the dwarf planets Eris, Makemake and Haumea along with their moons. Other notably large TNOs are Sedna, 2007OR10, Quaoar and Orcus.

On 14 July 2015, the 'New Horizons' probe flew 12,500 km above the surface of Pluto and sent a high resolution image of the TNO back to Earth.

The 'New Horizons' probe image of Pluto (above)

An artist's impression of the 'New Horizons' probe's flyby of Pluto (left)

Global Warming

Specification:

- Describe how the effects of global warming on Earth can be visualised by observing Venus' carbon dioxide rich atmosphere

Human activity and natural events are increasing the concentration of greenhouse gases such as carbon dioxide in the Earth's atmosphere. The increase in greenhouse gas concentration is believed to be causing a gradual increase in the average global temperature and may eventually lead to a change in the Earth's climate.

The effect of increased levels of carbon dioxide can be observed on Venus. The planet has a very dense CO_2 rich atmosphere which makes Venus' average surface temperature around 452°C, compared to around 14°C on Earth.

If the concentration of carbon dioxide in the Earth's atmosphere continues to rise, it will start to become like that of Venus and Earth's climate will change drastically, making the planet uninhabitable for humans.

An artist's impression of Venus and an industrial site in China which releases large amounts of carbon dioxide gas into the atmosphere

Space Probes

Specification:

- Describe the variety of space probes that space exploration agencies use to gain a better understanding of the Universe

Unmanned space probes have been an invaluable source of data and information about our Solar System and beyond. The main advantage of unmanned space probes over manned missions is that they do not have to support life and can travel with only essential equipment for thousands of years, taking readings whilst sunlight is still available.

The type of equipment on board an unmanned space probe depends on the particular mission and the type of information to be collected.

Data gathering tools

Spectrometers can record the chemical composition of planets as well as their related systems such as moons and rings.

Rovers can explore the surfaces of extraterrestrial bodies, examine ground composition and perform a variety of other functions.

Landers: Space probes often have a separate lander which can land on an extraterrestrial body like a moon or an asteroid and perform a number of scientific experiments as well as taking pictures.

Missions

The Cassini-Huygens probe was launched in 1997 and travelled to Saturn to investigate the elemental composition of the planet and its rings. The Huygens lander separated from the orbiter to investigate Titan by measuring atmospheric composition and recording images and sounds.

Voyager 1 and Voyager 2 were launched in 1977 to take advantage of the close alignment of Jupiter, Saturn, Uranus and Neptune and measured their weather, magnetic fields and ring systems.

Manned Missions

Specification:

- Evaluate the benefits and drawbacks of manned space missions

Benefits of Manned Missions

- Humans are better at problem solving and so provide more flexibility if something goes wrong

- Humans have to be trained, but do not cost as much as developing an intelligent computer system to run systems and solve problems

Limitations of Manned Missions

- Resources for life support have to be constantly supplied to keep the crew alive and healthy

- If the mission fails then human life is lost as opposed to just equipment

- It takes a long time and a lot of money to train an astronaut

- Astronauts who spend a long time in space suffer unpleasant medical conditions upon their return like muscle deterioration, mineral deficiency and cognitive problems

Although no-one has ever attempted a manned mission beyond our Moon, the International Space Station is constantly inhabited by a changing roster of astronauts

Natural Satellites and Rings

Specification:

- *Describe the origins and nature of the Martian and Neptunian satellite systems*

- *Describe the nature and composition of ring systems*

Martian moons: Phobos and Deimos

Phobos and Deimos are thought to have once been free-moving asteroids that were captured by Mars' gravitational pull. They both have irregular shapes and elemental composition similar to that of an asteroid.

Neptune's Moon: Triton

A picture of Phobos

One of Neptune's three moons, Triton is thought to be a captured object because it is relatively large, has a very high mass and rotates the opposite way to Neptune.

Rings

Saturn is famous for its ring system, but the other gas giants Jupiter, Uranus and Neptune also have rings. They are kept in place by the large planets' gravity and, in some cases, by shepherd moons. The composition of the ring particles varies; Saturn's rings are almost entirely composed of water ice.

2.2

Comets
&
Meteors

Astronomical Objects

Object	Where they are	Composition
Comet	In orbit around the Sun in the furthest reaches of the Solar System, fewer in number than asteroids	Ice and rock*
Asteroid	In orbit around the Sun, large numbers between Mars and Jupiter, variety of different sizes	Rock*
Centaur	Large bodies similar in size to asteroids but more similar in composition to comets, found in the outer Solar System, mainly between Jupiter and Neptune	Ice and rock*
Meteoroid	Small particles of asteroids and comets	Rock* and metals eg iron, nickel etc.
Meteor	An asteroid that has entered the atmosphere	Rock* and metals eg iron, nickel etc.
Meteorite	A meteor that has landed on Earth	Rock* and metals eg iron, nickel etc.
Micrometeorite	A very small meteorite	Rock* and metals eg iron, nickel etc.

The term 'rock' suffices for general descriptions but is heavily simplified and can refer to siliceous and carbonaceous compositions amongst others.

Comets

Specification:

- Label the nucleus, coma, dust and ion tails on a comet diagram

Comets are bodies made out of ice and rock. As they approach the Sun, some of the ice evaporates forming a coma of ice and dust around the nucleus. When the comet gets closer to a sun, two trails form in the opposite direction.

A curved dust tail forms which can be several million kilometres long. It forms as radiation from a sun impacts the comet. The tail contains ice particles which reflect sunlight and so appear bright.

A second tail made out of ions also forms. It is much straighter and forms as solar wind impacts the comet causing chemical changes. The fluorescent blue colour comes from the ionised elements.

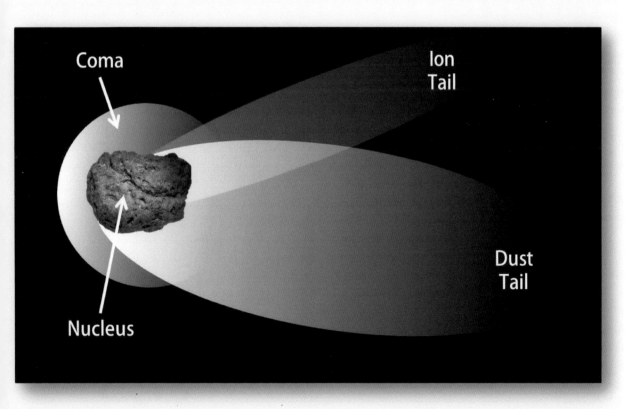

Orbit Shapes

Specification:

- Compare the orbits of planets and comets in terms of their shape and inclination to the ecliptic

Planets orbit the Sun in a prograde (anti-clockwise) motion when imagined from above, however comets orbit the Sun in a retrograde (clockwise) motion

Comets are also different from planets in that they have very inclined orbits, meaning that they do not orbit on the ecliptic, but at an angle to it so they can be observed from a wider range of positions.

All astronomical objects orbit in the shape of an ellipse, but whereas the orbit of a planet has very low eccentricity, meaning that it appears almost circular, a comet has an extremely eccentric orbit which forms an oval shape around the Sun. Some even have open parabolic orbits, meaning that they will not return to the same position.

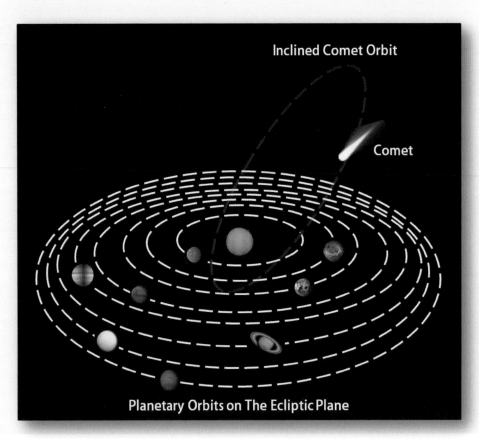

Belts and Clouds

Specification:

- Describe the origin and nature of the Kuiper Belt and Oort Cloud and explain how they are related to comets

- Explain how astronomers came to hypothesize the existence of the Oort Cloud

Kuiper Belt

The Kuiper Belt (named after Gerard Kuiper) is an area in the solar system between 100 and 1000 AU where short period comets originate from. A short period comet is one with an orbital period of under 200 years. The region is beyond the orbit of Neptune and is regularly intersected by Pluto (Pluto is now classed as a Kuiper belt object). The dwarf planets Haumea and Makemake also reside in the Kuiper Belt.

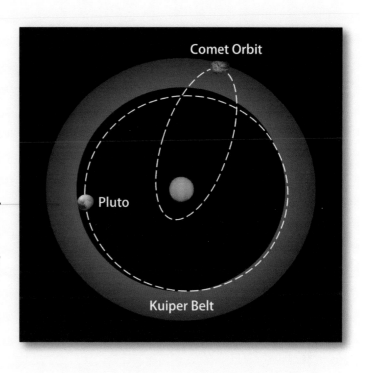

Oort Cloud

Studies of long-period comets suggest that there is a spherical region between 10,000 and 20,000 AU. Its existence was first hypothesised by Jan Oort who claimed that comets in the cloud would occasionally be disturbed by a passing star and nudged towards our Sun.

Without the existence of a structure like the Oort Cloud, long period comets with an orbital period of over 200 years would run out of ice and other material along their orbit.

Centaurs

Specification:

- Describe the nature and location of centaurs

Centaurs are astronomical objects with features similar to both asteroids and comets. Their orbits are reasonably eccentric ellipses and they are thought not to be permanent, with most centaurs only having been in their current orbit for a relatively short period of time.

Some centaurs are in orbit between Jupiter and Neptune, but the majority are between Saturn and Uranus.

The first centaur to be discovered is named 2060 Chiron and possesses a double ring system of its own.

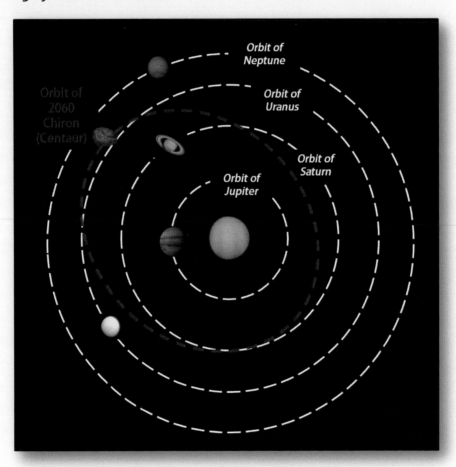

Asteroids

Specification:

- Describe the nature and location of asteroids

The majority of asteroids consist of rock, iron and silicate. Most asteroids in our Solar System orbit in a belt between Mars and Jupiter. This is roughly 2.2 to 3.3 astronomical units away from our Sun.

It is thought that they might be the remnants of an unformed planet, possibly due to gravitational influences from its surrounding planets.

The largest asteroid is called Ceres and was discovered in 1801. It has a diameter of about 945 kilometres. This is unusual, as only thirty asteroids have a diameter greater than 200 kilometres.

Vesta is the only asteroid that can be seen with the naked eye. It has an apparent magnitude of 5.6.

Photo of the asteroid Eros (top) Diagram showing the position of the asteroid belt (bottom)

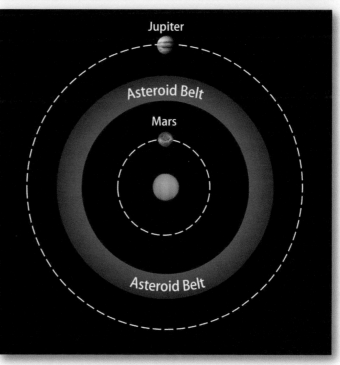

Meteor Naming

Specification:

- Describe the nature and origin of micrometeorites, meteorites and meteoroids

Cometary Meteoroid is a fragmented piece from a comet in space that was shed off a larger body.

Asteroidal Meteoroid is a fragmented piece from an asteroid in space that was shed off a larger body.

Meteor is a meteoroid that enters the Earth's atmosphere. Meteors travel at high speed, producing a lot of friction which makes them appear as a streak of light while they burn up.

Meteorite is a meteoroid that has passed through the Earth's atmosphere and impacted on the surface of the Earth.

Micro Meteorite is a meteorite with a diameter less than one micrometer. They are too small to be burnt up in the Earth's atmosphere.

Meteoroid (in Space)

Meteor (in Atmosphere)

Meteorite (on Earth)

Meteor Shower

Specification:

- Distinguish between sporadic meteors and meteor showers

- Explain the occurrence of meteor showers, using a diagram if necessary describe how they appear to deviate from a radiant point

Sporadic meteors may occur at any time as they are associated with asteroids or comets which do not follow a regular orbit.

Meteor showers are much more predictable as they are associated with comets or asteroids which follow a regular orbit.

Meteor showers are large numbers of meteoroids burning up in the Earth's atmosphere during a short period.

Meteor showers are most often seen after midnight as in that position, the Earth is facing forward into the trail of a comet or asteroid.

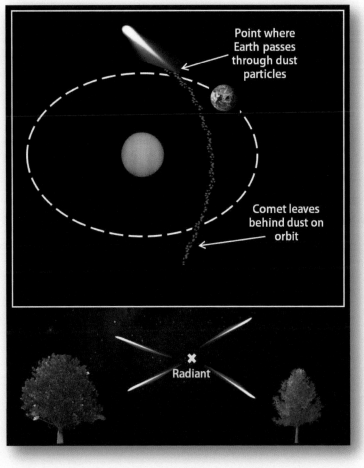

During a meteor shower, all meteors will appear to diverge from one point in the sky called the radiant.

A meteor with an apparent magnitude brighter than -3 is known as a fireball.

Impacts and PHOs

Specification:

- Describe the damaging effect of a collision between a PHO and the Earth

- Describe how PHOs are identified by amateurs and space agencies

- Explain why some PHOs are difficult to track

In 1994, Eugene Shoemaker and David Levy observed 21 fragments from a comet impact Jupiter during one day. Each impact caused huge explosions; scars were clearly visible on the surface for a long time afterwards. If this impact was replicated on Earth, a mass extinction event would follow, similar to those scientists believe occurred 65 and 250 million years ago.

Near Earth asteroids and comets are referred to as Potentially Hazardous Objects (PHOs) because their orbit takes them close enough to the Earth that collision is a real possibility, which would cause huge damage.

Astronomers currently have difficulty detecting comets on a trajectory towards us from the Sun; if scientists miss these or other kinds of PHOs then there is a distinct danger that a mass extinction could occur.

Currently the best way to protect ourselves from the danger of PHOs is to continually monitor near Earth objects with the hope that if a PHO is detected, there will be time to formulate a response.

2.3

Astronomers
Kepler's Laws
Calculations
Discoveries
Gravity

Solar System Discoveries

Astronomers

Specification:

- Describe the discoveries of the astronomers Copernicus, Tycho and Kepler and their effect on our understanding of the universe

Nicolaus Copernicus was the first astronomer to suggest a heliocentric model system which places the Sun in the middle of our Solar System with the Earth orbiting around it, rather than the previously accepted geocentric model which had the Earth in the centre. However due to the power of the Church, Copernicus' ideas were not at first accepted.

Tycho Brahe made several naked-eye observations, using basic instruments, which proved to be extremely accurate and were used in calculations by astronomers such as Kepler.

A statue of Nicolaus Copernicus

Johannes Kepler is best known for his three astronomical laws:

1. Planets move in elliptical orbits around the Sun

2. Planets move faster when they are closer to the Sun

3. There is a quantifiable relationship between the distance of a planet from the Sun and the planet's orbital speed

Discoveries

Specification:

- Describe Galileo's major astronomical discoveries

- Describe the process by which Ceres, Uranus, Neptune and Pluto were discovered along with key dates and astronomers

Galileo Galilei was the first astronomer to use a telescope. He observed that Venus had changing phases like the Moon due to its changing position in orbit around the Sun. He was also able to use his telescope to observe the Moon, seeing craters and mountain ranges in great detail for the first time. Additionally he observed the four moons of Jupiter which have since been called the Galilean Moons.

Uranus was first discovered to be a planet by William Herschel in 1781 using a telescope. He first wrongly described it as a comet, but with further observations, the object's lack of a tail and circular orbit confirmed it as a planet.

Ceres was discovered by Giuseppe Piazzi in 1801 and was at first wrongly identified as a star. Further observations of its movement and a new mathematical formula confirmed it to be an asteroid, the first asteroid to be discovered.

Neptune was discovered in 1846 by Johann Gottfried Galle and Heinrich Louis D'Arrest after having its position mathematically predicted by Urbain Le Verrier based on the gravitational influence a large body was having on the orbit of Uranus.

Pluto was first predicted in 1906 by Percival Lowell, but only observed in 1930 by Clyde Tombaugh. He used a blink comparator which compares the night sky and makes it easy to see which objects have moved. Pluto could be seen moving in front of a background of very distant stars which remained static.

Gravity

Specification:

- Describe the physical mechanics of Newton's inverse square law in relation to forces at work in the Universe

Isaac Newton was the first physicist to formulate an inverse square law for the relationship between the strength of gravitational force and the distance between bodies which the forces act on.

The gravitational inverse square law works very similarly to the dispersal of light: as the distance between two object increases, the gravitational forces between those two forces decreases by the square of the distance.

e.g. If a spacecraft moves 2 AU away from Earth, the Earth's gravitational influence on the craft would decrease by a factor of $2^2 = 4$

Newton also worked on deriving Kepler's laws and using his theory, was able to account for the shape of the Earth, the existence of tides and why the moon is tidally locked to Earth amongst other things.

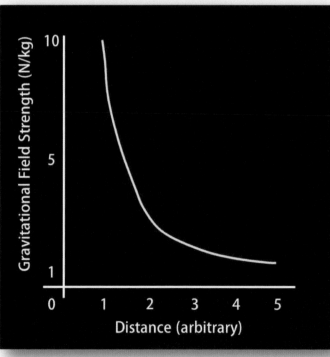

2.4

Exoplanets
Goldilocks Zone
ET Life
ET Water
ET Communication

Exoplanets

Exoplanets

Specification:

- Explain why it is difficult to detect exoplanets visually

- Describe the methods, with a diagram necessary, used to detect exoplanets including astrometry, transit detection and radial velocity detection

Directly observing exoplanets is difficult as they appear very faint and are positioned very closely to their star, however other methods are more viable.

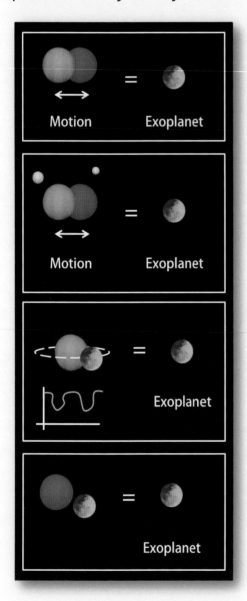

Radial Velocity Detection

Exoplanets exert a gravitational pull on the stars in their systems which creates 'wobble'. The wobble creates a measurable shift in the frequency of light. A shift in the frequency of light from some distant solar systems has been detected, indicating the presence of exoplanets.

Astrometry Detection

This technique uses the same principle that stars with exoplanets have a wobble, but instead of direct observation of the star, movement of the star compared with the background is detected.

Transit Method Detection

When observing the light intensity of a star over a period of time, occasionally a dip is visible which signifies an object moving in front of it. If the dip is of a constant intensity and at regular intervals then it is likely to be an exoplanet in orbit around the star.

Optical Detection

In certain circumstances exoplanets can be detected optically if bright, distracting sources of light from stars is blocked out

Goldilocks Zone

Specification:

- Describe the mechanics of the 'goldilocks zone' and why it is significant in relation to finding intelligent life outside our Solar System

The Goldilocks Zone, more commonly known as the Habitable Zone, is the area around a star in which life could potentially exist. A planet is deemed suitable to sustain life if liquid water is present on the surface (or there are the correct conditions for its formation) and it has a suitable atmosphere and a magnetosphere with the ability to hold water on the surface.

The Goldilocks Zone around a star can be fundamental to astronomers in their search for life, as it enables them to focus their search. However, this ignores the possibility of life being present on a moon or beneath the surface of a planet outside of the habitable zone.

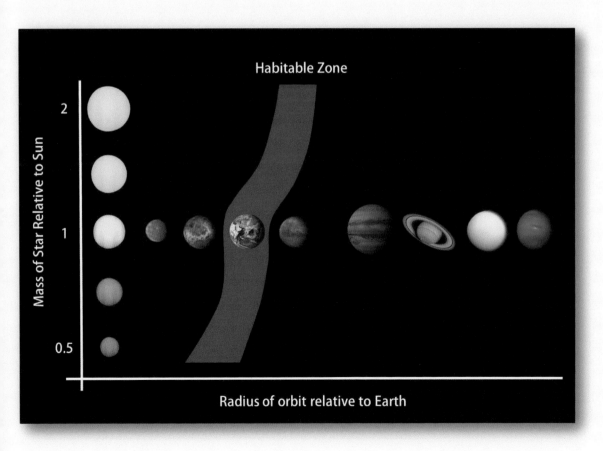

Extraterrestrial Life

Specification:

- Describe how space agencies and amateurs are attempting to make contact with extraterrestrial life outside of Earth

The general consensus among modern scientists and astronomers is that life must exist elsewhere and in the majority of cases would be carbon-based. Furthermore, the life form would probably require water for its survival and formation.

The Search for Extraterrestrial Intelligence Institute (SETI) was founded in 1984 with a mission to 'explore, understand, and explain the origin and nature of life in the universe'. They have a network of thousands of satellite dishes and computers with which they analyse signals collected from space in the hopes of intercepting alien communications which could be deciphered.

The Voyager space probes both had a golden record attached to them which contained images and sounds about life on Earth. Voyager 1 will pass within 1.6 light years of a star and so it is hoped that an advanced civilisation with high speed spacecraft will be able to find and use the record to make contact with Earth.

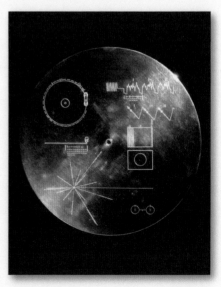

Extraterrestrial Water

Specification:

- Explain why water is currently considered a requirement for the formation of life

- Compare the validity of the two main theories as to how liquid water formed on Earth

- Describe some of the techniques used by space agencies to find evidence to prove the condensation theory or the delivery method

It is currently thought that water is required for life to exist due to its property as a solvent which allows a variety of essential chemical reactions to occur. The origins of liquid water on the Earth are unknown but the two leading theories are the condensation theory and the delivery theory, both of which have evidence to support them.

Condensation theory:

This hypothesis claims that water has always been present on Earth, but it was initially within the terrestrial body of the planet. It is thought that as the Earth cooled, water vapour escaped from the surface into the newly formed atmosphere and then condensed into liquid form and fell to the surface.

Delivery theory:

Because of the highly cratered appearance of the Moon, it is believed that the Earth underwent a period of heavy bombardment during the early stages of its life. The collisions were probably from small celestial bodies such as comets which contain large amounts of ice. During a high energy collision, the ice would have melted and formed liquid water which over time could have fed into the oceans. The likelihood of all the water in the oceans coming from comets is slim, as it would take a sizeable number of collisions to produce the water required to cover over 70% of the Earth's surface.

Extraterrestrial Communication

Specification:

- Compare the advantages and disadvantages of communicating with extraterrestrial life

The advantages and disadvantages of communicating with extraterrestrial life has long been a topic of debate between scientists who have differing opinions about whether it is a good idea or not . It is strongly believed that intelligent life exists outside of our Solar System and every new exoplanet discovered by astronomers makes the possibility of contacting an alien civilisation more likely.

Benefits:
- Alien forms of life could provide insights into cures for ill-health.
- Knowledge unknown or forgotten by humankind may be shared by other intelligent life forms. This may allow for advancement in technological understanding and its technical application.

Drawbacks:
- The discovery of new life opens up a vulnerability to new diseases.
- Extraterrestrial life may have aggressive intentions, which could lead to the mass extinction of Earth's native species.

Photograph showing a radio telescope searching for extraterrestrial life

Topic 3

Stars

Topic 3 Overview

Topic 3 is a more practical overview of the Universe which provides information about the features of the night sky including constellations, galaxies and nebulae. The second half of the unit then focuses more on the physical mechanics of observable astronomical objects and ends with a comprehensive walkthrough of the stages involved with a star's life cycle.

Topic 3.1 - Constellations: A detailed account of the various observable objects in the night sky as well as diagrams of the most recognisable constellations and asterisms.

Topic 3.2 - Observing the sky: A short chapter on factors and conditions which have to be considered when undertaking practical observation of the sky. The unit also includes notes on startrail recording and the Messier catalogue.

Topic 3.3 - Stars: A more detailed study of the physical mechanics going on inside stars and their systems. The unit then covers more recent astronomical discoveries such as parallax distance measurements and Cepheid variable stars.

Topic 3.4 - Evolution of Stars: An in-depth walkthrough of the various stages of a star's life cycle from its conception to its eventual death. The last part of the unit is a more hypothetical investigation of what astronomers believe about black holes.

Key Facts and Figures

Formula to calculate the apparent magnitude of an object

$M = m + 5 - 5(\text{Log}_{10}d)$

(M = absolute magnitude) (m = apparent magnitude) (d = distance in parsecs)

Formula to calculate the rotational period of the Earth from startrails

$T = (360 \div \theta) \times t$

(T = rotational period of Earth) (t = time over which the recording occurred)

Spectral Classes (from highest to lowest temperature)

O, B, A, F, G, K, M

3.1

Stars
Constellations
Pointers

Constellations

Appearance of Stars

Specification:

- Define and describe the appearance of stars, double stars, asterisms, constellations, open clusters, globular clusters and nebulae

Star: a bright sphere of plasma whose colour and size can vary depending on its temperature and distance from Earth.

Double Star: to an observer a double star appears as a binary star system, however, two separate stars are present and only appear close when observed from Earth.

Asterism: a pattern of stars within the night sky. An asterism can either stand alone or be linked to a wider constellation. A well-known example of an asterism is the Plough.

Constellation: a pattern of stars which divides up the night sky. In total there are 88 constellations in the Northern Hemisphere. Examples include Orion, the Hunter and Cygnus, the Swan.

Open Cluster: a collection of stars located close together. They do not form recognisable patterns, unlike asterisms and constellations.

Globular Cluster: a group of stars forming a spherical shape which appears as a fuzzy ball in the sky. In the majority of cases the stars themselves are red giants and gather in the central region of the cluster.

Nebula: a blurry band across a section of the night sky. Most visible nebulae are found within our own galaxy.

Constellations

Specification:

- *Describe how stars in a constellation are labelled using the Greek alphabet and be able to recall the alphabet from alpha to theta in order*

- *Explain how the modern list of constellations was established and the cultural factors involved*

The luminosity of stars in a constellation varies so it is possible to distinguish between them. The current labelling system, consisting of the Greek letters α to ε, was invented by Johann Bayer in 1603. The letter α represents the brightest star within the constellation while the letter ε is designated to the dimmest star. The letter is a representation of the star's relative magnitude compared to other stars in the constellation, not the star's absolute magnitude. In addition to identifying individual stars there is an official modern list of constellations. This was agreed by the International Astronomical Union (IAU) in 1922. The naming of constellations was first introduced by the ancient Greeks and, as a result, Greek mythical references are often the source of constellations' names. Throughout history, different civilizations have assigned different names to the constellations, relating them to their own culture.

Name	Symbol
Alpha	α
Beta	β
Gamma	γ
Delta	δ
Epsilon	ε
Zeta	ζ
Eta	η
Theta	θ

Pointer Stars

Pointer stars are used as arrows to enable easy location of other celestial objects in the night sky. This technique for locating objects is relatively simple and involves drawing a line from a set of easily identifiable stars to a harder to find location (see diagrams). A well-known use of this technique is using stars within the Plough to locate Polaris.

To an observer at a given latitude, the visible constellations change during the year with only a few remaining visible at all times. This change occurs as the Earth orbits the Sun and rotates about its axis. The constellations that remain observable throughout the year are known as circumpolar constellations as they never set. Constellations are not visible during the day due to the selective scattering of wavelengths which gives the daytime sky a blue, red or orange appearance depending on the time of day and year.

The Plough

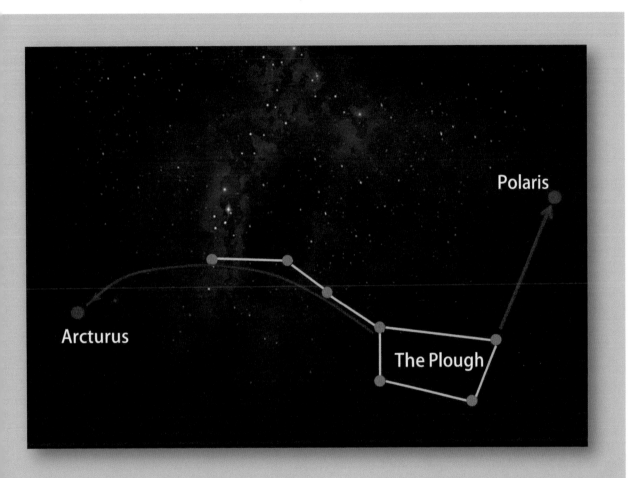

A diagram showing The Plough and the use of pointers. The Plough (the Big Dipper) is not a constellation but an asterism (a popular pattern of stars).

The Great Square of Pegasus

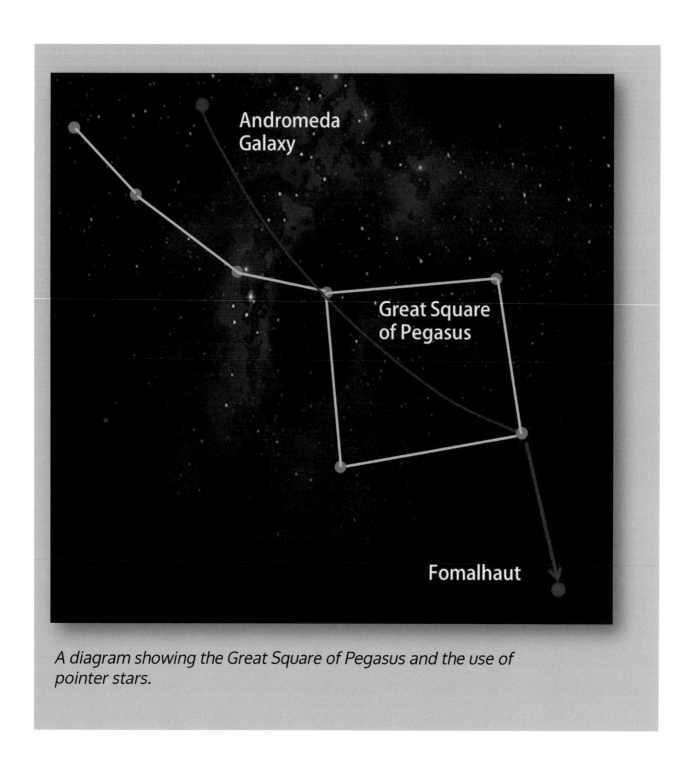

A diagram showing the Great Square of Pegasus and the use of pointer stars.

Orion (The Hunter)

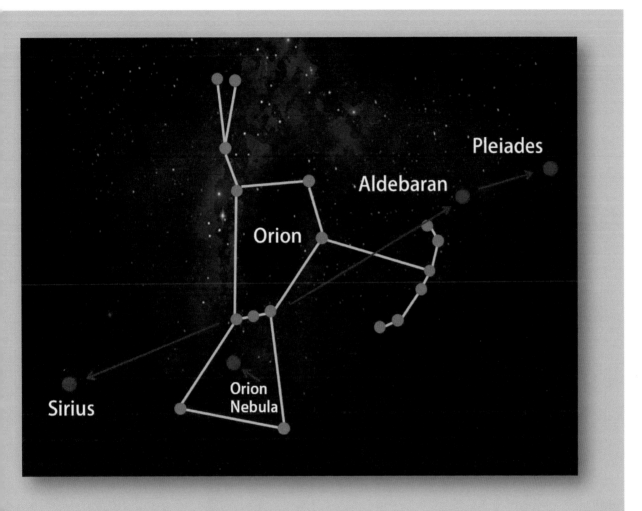

A diagram showing the constellation Orion and the use of pointer stars.

Cassiopeia

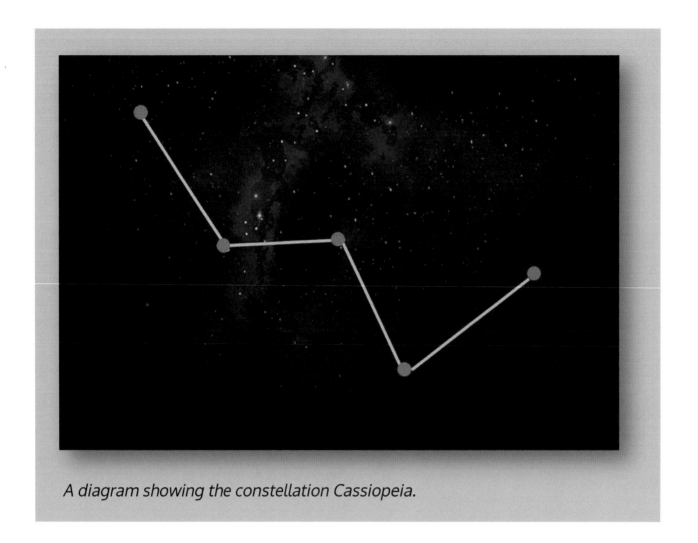

A diagram showing the constellation Cassiopeia.

Cygnus the Swan

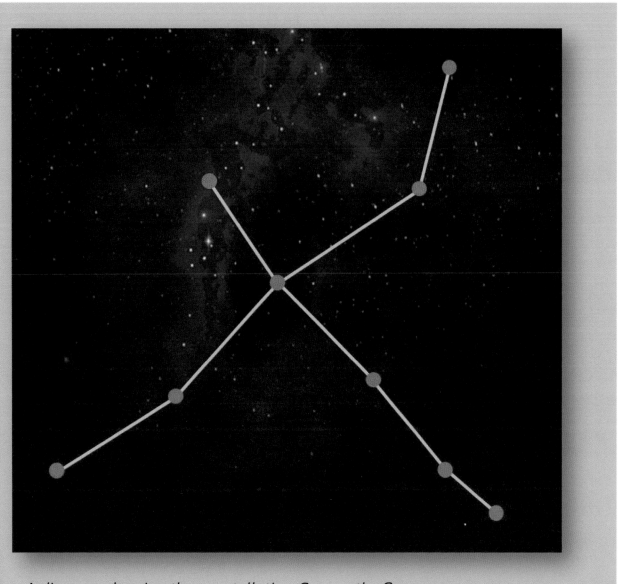

A diagram showing the constellation Cygnus the Swan.

3.2

Star Trails
The Night Sky
Terms
Messier Catalogue

Observing the Sky

Star Trails

- *Explain the apparent motion of stars while the Earth rotates*

- *Explain how long-exposure photos of the stars around Polaris can be used to calculate the rotational period of Erath*

Star trails are arcs of light on photographs captured using long-exposure photography which show the path of a star as it moved across the night sky. Star trails allow astronomers to accurately determine the rotational period of the Earth using the following equation:

Rotational period = $(360 \div \theta) \times t$

In this equation the symbol θ is assigned to the angle between the Pole Star and either end of the star trail while t is the time over which the photograph was taken. Using this method, the Earth's rotational period can be calculated to be 23 hours and 56 minutes.

The Earth's rotational period is less than 24 hours because the Earth rotates as it orbits around the Sun. The four-minute difference between the sidereal day and solar day arises from the Sun having to move an additional 1° across the sky. This is because a solar day is complete when Sun to returns to the same location within the sky.

Example:
A student takes a photograph with an exposure time of 60 minutes. The measured angle is 15°. Using this information, the rotational period of the Earth (T) can be calculated as follows:

Long exposure photograph

T = $(360 \div 15°) \times 60$ [15° is a rounded approximation, really it is around 15.0417...°)
    = 1436 minutes (23 hours 56 minutes)

The Night Sky

Stars in the night sky appear to move from east to west due to the counter-clockwise rotation of the Earth.

Many stars in the night sky are circumpolar which means that their circular motion in the night sky never brings them below the horizon and they are always observable. Polaris is a relatively bright star in the night sky which is circumpolar when observed from the northern hemisphere. Polaris is fixed, meaning that it has barely any motion and appears static, with all other stars rotating around it, which is useful when a central point in the sky is needed for taking star trail pictures.

Polaris is located close to latitude 90°, meaning that when observed from the North Pole, it appears directly above. Due to this phenomenon, the elevation of Polaris from the horizon when observed from any location is the same value as the latitude of that location.

The longer the human eye is kept in completely dark conditions during an observation period, the better the eyes become adapted to the dark. This adaptation allows observers to see much fainter stars up to magnitude 6 with the naked eye.

An observer's zenith is the point in the sky directly above the observer. The term culmination is used to describe the point at which a star crosses the zenith.

The time at which stars culminate when they are due south can be calculated by reading a star chart. The chart provides information regarding the star's latitude and longitude at specific times, as well as its culmination time on a specific day of the year.

The Messier Catalogue

Specification:

- Describe the origin, purpose and nature of the Messier Catalogue

The original Messier Catalogue was compiled by Charles Messier in 1771. He was interested in observing comets and created a list of 45 objects which could be mistakenly identified as comets. It remains an important reference and has been expanded to 110 objects, but nowadays digital recognition is widely used which replaces direct observation with reference to the Catalogue. The Catalogue contains a huge range of objects including quasars, blazars and nebulae.

The Andromeda Galaxy (left), the Hercules globular cluster (bottom left) and the Crab nebula (bottom right) are all examples of Messier objects

3.3

Stars

Binary Stars
Double Stars
Clusters
Magnitude
Distances
Light
Spectrums
Classes
HR Diagram

Binary Stars

Binary stars are often hard to visually identify, but a variety of techniques can be used to detect them. By measuring the light intensity of a system, a light curve can be produced and if a repeating pattern such as the one in the diagram below can be seen, the presence of a binary star system is likely.

The period of the binary star system's rotation can be found by measuring the time taken for one complete waveform on the diagram. During one complete cycle there are two drops in light intensity, one of which is greater than the other. The greater drop occurs during the transit of the smaller star in front of the large star which is more luminous. The diagram below depicts one full cycle.

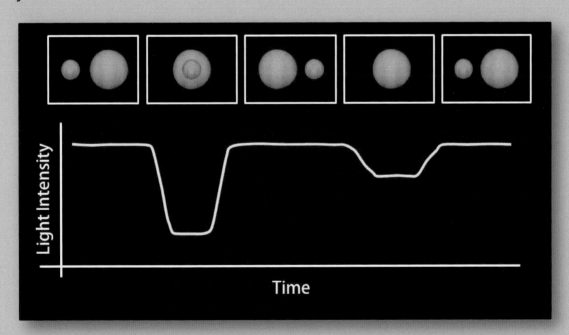

Graph showing the changes in light intensity during the orbital period of a binary star

Double Stars

- Compare the nature or optical double stars and binary stars

Optical double stars are a pair of stars that falsely appear to be in a binary star system when viewed through a telescope without adequate magnification. Optical doubles are in fact two stars which are in line when viewed from Earth, but are actually separated by a great distance.

A **binary star system** is a star system consisting of two stars orbiting around a common centre of gravity. Systems consisting of two, three, four or more stars are called multiple star systems.

When observing, it can be hard to distinguish between optical double stars and a binary star system. When observing a binary star, a small wobble can be seen due to the gravitational forces they exert on each other.

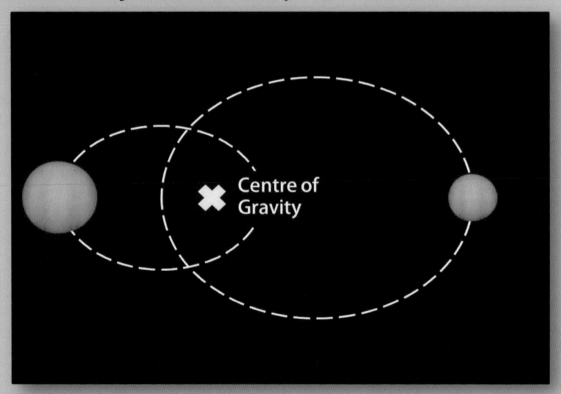

A diagram showing a binary star system

Clusters

A star cluster is a collection of several stars which are gravitationally bound together. This means that the stars exert a gravitational pull on one another and cannot escape.

Star clusters can be further categorized into two distinct types:

- **Open clusters** contain far fewer stars, most of which are far younger. Stars within open clusters are spaced much further apart as the gravitational forces between them are weaker.

- **Globular clusters** are recognisable by the hundreds of thousands of ancient stars that they contain which are strongly bound together by gravitational forces.

Enhanced images of an open cluster (left) and a globular cluster (right)

Magnitude

Specification:

- *Define the terms absolute magnitude and apparent magnitude*

- *Understand the mechanics of the magnitude scale and how including the direction of the scale and the intervals between integer values*

- *Recall the equation to calculate absolute magnitude from apparent magnitude and distance values provided and be able to rearrange it according to the question*

Apparent magnitude is defined as the measured luminosity of a celestial object as observed from Earth. In comparison, absolute magnitude is defined as the measured luminosity of a celestial object as if it were ten parsecs from Earth.

The magnitude scale is a numerical scale where the lower the magnitude, the brighter the object. The dimmest object that the naked human eye can see is around 6 on the scale while our Sun is around -27. Each increase in the order of magnitude represents an object becoming 2.5 times dimmer.

The magnitude formula contains three variables: M (absolute magnitude), m (apparent magnitude) and d (distance in parsecs).

The formula can be rearranged to allow for the calculation of any of these three values, however it is most commonly arranged to find the absolute magnitude.

$$M = m + 5 - 5(\text{Log}_{10}d)$$

M = absolute magnitude, m = apparent magnitude, d = distance in parsecs

Distances

- Understand the use of parsecs in calculations

- Explain how parallaxes are used to calculate distances

The parallax phenomenon in physics causes the apparent position of a nearby star (no more than 100 parsecs away) to shift with respect to the background of distant stars throughout the course of a year. The angular shift can then be measured and trigonometry can be used to determine distance.

This use of parallaxes to measure distances ceases to work after 100 parsecs because the angular shift becomes too minute to measure. Over greater distances other methods are used.

A parsec is a unit of distance equal to 3.26 light years. Astronomers use very small angles when using the parallax method so minutes and seconds, instead of degrees, are used to measure angular values. A minute is a 60th of a degree and a second is a 60th of a minute. If the angle p (in the diagram) created between the radius of the Earth's orbit and a star is 1 arc second (3600th of a degree), then the star is 1 parsec away.

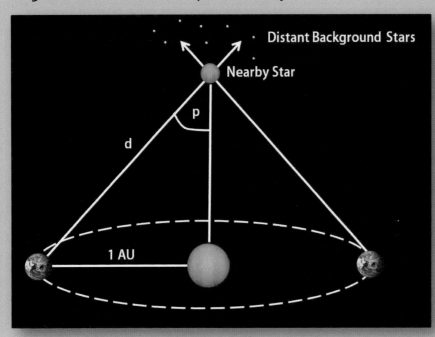

A diagram showing the method of heliocentric parallax to determine the distance to nearby stars

Light Dispersal

As light travels, its intensity lessens. This occurs because of the increased area it covers. The area in which it spreads is the square of the distance it travels. The intensity of the light at a given point is proportional to $1/r^2$, where r is the distance from the source.

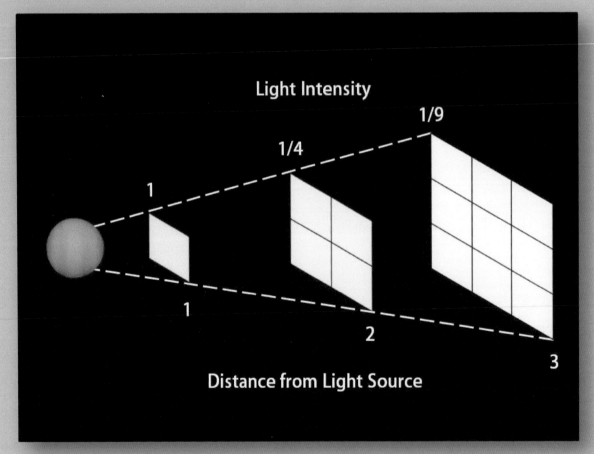

A diagram aiding the explanation of the inverse square law nature of the intensity of light

Cepheid Variable Stars

A Cepheid variable is a star whose luminosity varies over a period of time in a regular cycle. A Cepheid variable star can be identified by observation of its light curve which has a regular and repeating pattern of luminosity where the period of the star is the time between two consecutive peaks in magnitude.

After calculating the time period, the data can be used to determine the star's absolute magnitude (M). This in turn allows the distance (d) from Earth to be calculated using the following equation: $M = m + 5 - 5(\text{Log}_{10}d)$. In the equation m is apparent magnitude, M is absolute magnitude and d is distance.

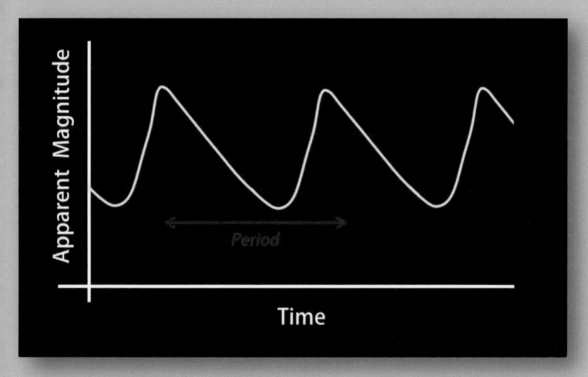

A Cepheid Variable light curve

Spectrums

Specification:

- Describe how temperature, elementary composition, mass, variability, velocity, distance and age can be determined from a spectrum of light

A telescope attached to a spectrometer which captures and splits light from celestial bodies can be used to provide data that can be analysed to reveal:

- Temperature
- Elemental composition
- Mass
- Variability
- Velocity (Doppler shift)
- Distance from Earth
- Age
- Absorption Spectra
- Emission Spectra

The stellar spectrum is a gradation of colours interspersed with black lines and can be interpreted to enable the individual properties of the star to be determined. The colours found within the spectrum indicate the elementary composition of the star and range from red to blue in the visible light spectrum, but can also go beyond. From this, the stage of the star in its lifecycle can be determined. The stellar spectrum contains dark absorption lines and white emission lines. The position and type of the line indicate whether a certain type of atom absorbs or emits light. The position of these lines allows astronomers to produce an accurate estimate of the star's age. For example, if the absorption lines are found within the hydrogen section and emission lines are found throughout the helium section, it is clear that the star has already begun to move along the main sequence as it has already fused all the hydrogen present in the beginning. The velocity of the star and its distance from Earth can be determined by taking a number of recordings over a period of time and comparing them side by side. By observing the change in wavelength, red or blueshift can be found which can then be used to calculate the star's velocity and distance from Earth.

Star Classification

During direct visual observations, most stars appear white in colour, while a small number appear to be different colours. Stars actually emit a range of different colours and it is because our eyes are unable to distinguish between colours in low light conditions that we see the majority of stars as white.

In addition to the range of wavelengths that each star emits, properties such as size, temperature and composition can be estimated from its colour. By studying these factors, the age of a particular star can be determined because its properties change as it evolves.

Stars can be classified according to their spectra (colour and wavelength) and temperature. In total there are ten classes, of which seven are in the visible spectrum. These seven can be remembered using the following mnemonic: 'Oh, Be A Fine Girl Kiss Me'.

Class	Colour	Temperature
O	Blue	Above 31,000 K
B	Blue - White	9750 - 31,000 K
A	White	7100 - 9750 K
F	Yellow - White	5950 - 7100 K
G	Yellow	5250 - 5950 K
K	Orange	3800 - 5250 K
M	Red	2200 - 3800 K

Table showing the classification of stars where O is most luminous and M is least luminous

Hertzsprung Russell Diagrams

Specification:

- Be able to sketch and/or annotate a Hertzsprung Russell diagram

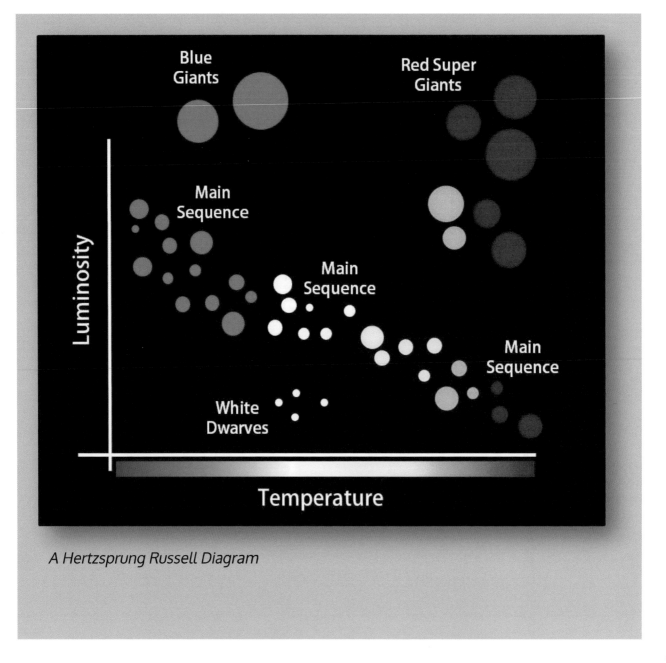

A Hertzsprung Russell Diagram

3.4

Evolution
Star Birth
Neutron Stars
Black Holes

Evolution
of Stars

Evolution of the Sun

Specification:

- Use specialist terms to describe the lifecycle of a star with 1 solar mass and give details about the five key stages

The life cycle of our Sun consists of five main stages. In each stage, the size, appearance and temperature changes.

Stage 1 - Formation: Not long after the beginning of the Milky Way, the Sun formed from a collection of gas and dust pulled together by gravity. During this time the Sun's internal pressure and temperature increased.

Stage 2 - Protostar: The Sun is now spherical in shape, however, its initial formation has not yet completed as contraction continues whilst the temperature in the core continues to rise.

Stage 3 - Main Sequence Star: The Sun has now finished forming and fusion has begun. This involves collisions between hydrogen atoms which creates energy (heat) and leads to the formation of a helium three atom. Stage 3 occurs for a duration of about 10 billion years and during this time the Sun moves along the main sequence.

Stage 4 - Red Giant: This stage begins as hydrogen in the core becomes depleted and hydrogen starts to fuse in the outer layers. As this occurs the star expands and subsequently becomes a Red Giant.

Stage 5 - White Dwarf: Fusion ceases completely causing the outer layers of the star to be ejected, which later form planetary nebulae. After this has occurred, a very hot, dense White Dwarf Star is left.

Star Life Cycle

- Use terms such as emission nebulae, absorption nebulae, open clusters, planetary nebulae and supernovae to describe the lifecycle of a star

Star Birth

Star formation occurs when gas and dust condense into a sphere due to gravity which leads to a rapidly spinning accretion disc. As the disc gathers material its temperature increases and this in turn starts a fusion chain reaction. In some cases, a Brown Dwarf forms due to a lack of matter.

- **Open Clusters** are groups of a few thousand young stars which are gravitationally bound together. All the stars found within the cluster come from the same molecular cloud.

- **Emission Nebulae** are clouds which are made from high temperature gas. Stars within or near the nebulae heat the gas as they emit large quantities of ultraviolet radiation. This in turn causes electrons to ionise which leads to the emission of electromagnetic radiation.

- **Absorption Nebulae**, otherwise known as Dark Nebulae, appear as dark patches in the sky as a result of gas and dust collecting to form a cloud so that light emitted from distant stars is blocked from the observer. Absorption Nebulae can be seen when viewed in front of the Milky Way.

Star Death

Star death occurs when a star reaches the end of the main sequence, as seen on the Hertzsprung-Russell Diagram. The final outcome is dependent on the star's mass at the time of death.

- **Planetary Nebulae** are ring-shaped nebulae. Their formation occurs as shells of gas expand out from an aging star. A planetary nebula occurs if the star had a solar mass of less than four.

- A **Supernova** appears after a star suddenly increase in brightness. This phenomenon occurs due to an explosion that ejects the majority of the star's mass. During the death of a star, a supernova will normally occur if the star had a solar mass greater than four times that of our Sun, however this is dependant on elemental composition.

Star Life Cycle

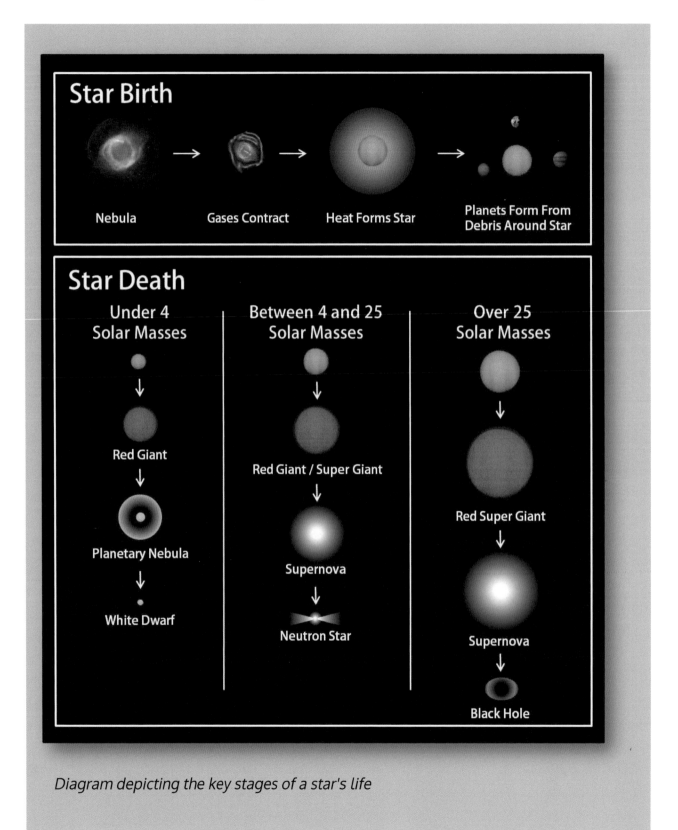

Diagram depicting the key stages of a star's life

Neutron Star

A neutron star occurs as a consequence of a red giant exploding towards the end of its lifecycle. This explosion creates a supernova and leaves behind a neutron star the size of Earth.

This extreme condensation of matter creates a very dense star which results in the whole core being composed of neutrons. The core is not solid, but liquid in form and has a typical mass that is 1.5 times greater that of our Sun while having a diameter of only 12 miles.

During the formation of a neutron star it rotates very quickly, often completing full rotations in seconds. The formation of a neutron star can be detected as large quantities of radio pulses are emitted as it begins to spin, giving rise to the name pulsar.

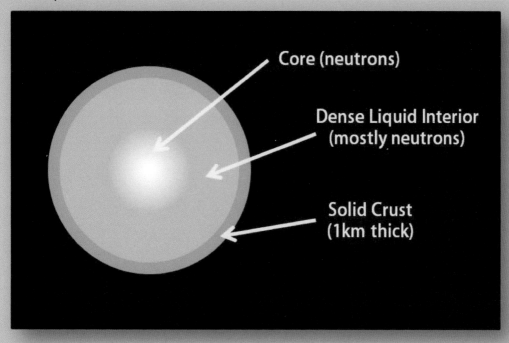

Core (neutrons)

Dense Liquid Interior
(mostly neutrons)

Solid Crust
(1km thick)

A diagram of a neutron star

Black Holes

A black hole is formed when a large star explodes and all its mass collapses into a miniscule singularity. It has been impossible to observe a black hole directly with modern technology as it appears black against the black backdrop of space, however objects like Cygnus X-1 which are thought to be black holes have large disks of material surrounding them, so techniques like gravitational lensing allow astronomers to observe black holes indirectly.

Astronomers have predicted the existence of black holes by observing binary star systems where matter disappears whilst other material seems to be in a static orbit around nothing. This disk of orbiting matter known as an accretion disk spins very quickly and sends out measurable x-rays.

An artist's impression of a black hole

Topic 4

Galaxies and Cosmology

Topic 4 Overview

Topic 4 is a more sweeping view of galaxies and the Universe as a whole. The unit begins with a detailed case study of the Milky Way, but then expands its scope to other galaxies further afield. An introduction to the observation and classification of galaxies is then given along with a further breakdown of the types of galaxies. Topic 4 then concludes with an overview of the physical mechanics at work in the Universe along with theories about its origin and future.

Topic 4.1 - The Milky Way: A short study of our Milky Way including details about its size, origin and nature, as well as a list of its key features.

Topic 4.2 - Galaxies: An overview of the observation and classification of galaxies, with further information about the types of active galactic nuclei and galactic grouping.

Topic 4.3 - Cosmology: A very theoretical chapter about the physical mechanics associated with cosmology such as the Doppler Effect and Hubble's law. The unit then ends with a collection of the most widely accepted hypotheses explaining the beginning and end of the Universe.

Key Facts and Figures

Number of Stars in the Milky Way

Anywhere from 100 billion to 400 billion stars

Diameter of the Milky Way

100,000 light years

Classification of the Milky Way

Sb class

Order of Organisation

Galaxy → Local Group → Cluster → Supercluster → Universe

4.1

The Milky Way

**Size and Shape
Features
Observation**

Observations

Specification:

- *Label a diagram of the Milky Way and locate key features such as the our Sun, galactic dust and globular clusters*

- *Explain the appearance of the Milky Way when observed from Earth*

- *Explain why 21 cm radio waves are used to observe the rotation of galaxies as opposed to visible light*

When observing the Milky Way from Earth it appears as a bright band of speckled white light across the dark background of the night sky. It is becoming increasingly difficult to view as it can only be seen in regions where there is little or no light pollution.

The actual appearance of the Milky Way varies depending on the direction of observation due to its disc-like shape. From above, the large central nucleus and outward-pointing spiral arms can be clearly seen, whilst from the side it appears relatively flat with a large central bulge.

In total, the Milky Way has a diameter of 100,000 light years and contains roughly 200 billion stars. Like a solar system, all objects rotate around a central nucleus, with the Sun taking approximately 250 million years to complete one full orbit as it is two-thirds of the galaxy away from the Centre. Astronomers wanting to determine the rotational period of galaxies use 21 cm radio waves which allow full observations to be made as interstellar cosmic dust and gas obscure parts of galaxies when making direct observations with visible light.

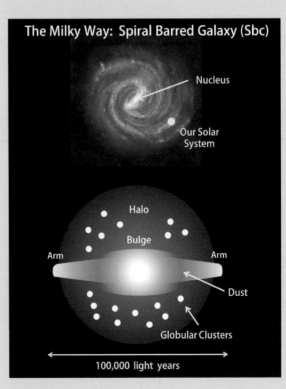

A diagram showing the shape and size of the Milky Way

4.2

Galaxies

Classifications
Tuning Fork
Radiation
Active Galaxies
Groups

Classification of Galaxies

Specification:

- Identify a spiral, barred spiral, elliptical or irregular galaxy from a picture or diagram

During aided observations, it is possible to distinguish between spiral, barred spiral, elliptical and irregular galaxies. The different classifications are shown in photographs below.

Spiral galaxy

Barred Spiral galaxy

Elliptical galaxy

Irregular galaxy

Hubble Tuning Fork Diagram

Specification:

- *Sketch and label a tuning fork diagram*

- *Label the place of the Milky Way on a tuning fork diagram*

The Hubble Tuning Fork Diagram was invented by the astronomer Edwin Hubble in 1926 as a means of easily classifying galaxies based on their appearance. It consists of three main categories: Elliptical (E), Spiral (S) and Barred Spiral (SB) and gives an indication of the degree of ellipticity and compactness of the spiral arms. The diagram allows astronomers to make quick and accurate classification while undertaking visual observations.

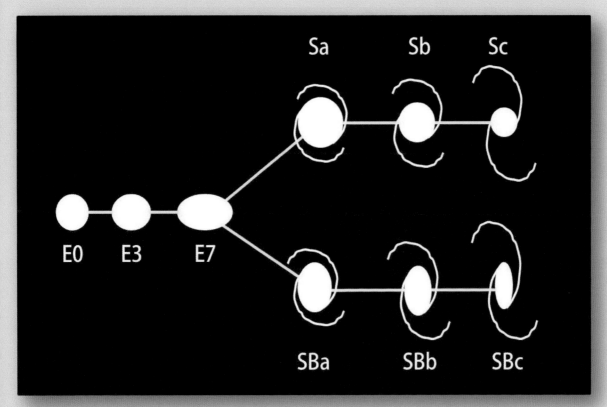

A diagram showing a Hubble's Tuning Fork. The Milky Way is an Sb type galaxy

Radiation

As with stars and nebulae, galaxies emit electromagnetic radiation that can be detected. Galaxies emit radiation with a variety of different wavelengths including radio waves, infrared, visible light, ultraviolet, x-rays and gamma rays. The type and quantity of emission depends on several factors including the age of the stars in the galaxy and their temperature and elemental composition.

An active galactic nucleus (AGN) is present if the area at the centre of a galaxy has a luminosity across the spectrum that is greater than average. Galaxies which fit these criteria are called active galaxies and they include quasars, blazars and Seyfert galaxies.

Astronomers believe that the formation of an AGN can be explained by the accretion of matter from a supermassive black hole. An accretion disc forms around the black hole which in turn produces enormous amounts of electromagnetic radiation across the entire spectrum.

Image showing the Eagle nebula, the home of several active star forming regions

Active Galaxies

Seyfert galaxies visually appear to be normal spiral galaxies, but their nuclei, which are normally about 10 light years in diameter, emit more radiation than the whole Milky Way, mainly in the form of infrared radiation. Many Seyfert galaxies are also weak radio wave emitters.

Blazar galaxies are observed through their blazar jets which point in the general direction of Earth and contain a variety of radio wavelengths. Their appearance varies greatly depending the angle of the jet relative to the observer. In some cases they emit x-rays as well as radio waves.

Quasars (quasi-stellar radio sources) are active galactic nuclei which emit so much light that it is impossible to observe the stars in the rest of the galaxy. Their name is misleading as it has now been found that only about 1 in 200 quasars emit radio sources, however every quasar emits electromagnetic radiation of some sort.

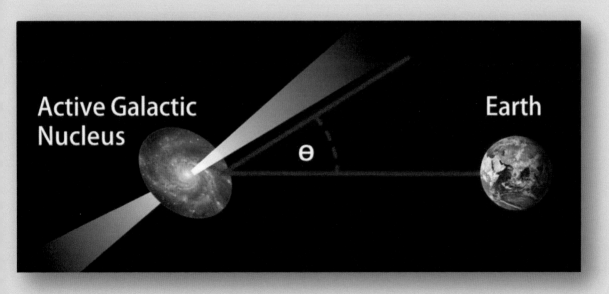

Galactic Grouping

Specification:

- Describe the nature of our local group of galaxies.

- Name some galaxies in our local group including the Large and Small Magellanic Clouds, the Andromeda Galaxy and the Triangulum Galaxy

- Describe how galaxies are organised into groups in increasing sizes from the smallest local groups to the largest superclusters in the Universe

The Milky Way is one of around 30 galaxies in our Local Group. Galaxies within the Local Group are gravitationally bound and when viewed from Earth they appear to move towards us, which we can detect as they show blue shift.

In addition to the Milky Way, other galaxies within our Local Group include the Large Magellanic Cloud, the Small Magellanic Cloud, the Andromeda Galaxy (M31) and the Triangulum Galaxy (M33).

The Large Magellanic Cloud is the third closest galaxy to the Milky Way and is believed to be a possible satellite of it. It is categorized as an irregular galaxy with a large central bar.

The Small Magellanic Cloud, a dwarf galaxy, is also an irregular galaxy and has a prominent bar through the centre.

The Andromeda Galaxy (M31) is a spiral galaxy like the Milky Way. In good conditions it can be observed with the naked eye.

The Triangulum Galaxy (M33) is also a spiral galaxy and is sometimes referred to as the pinwheel galaxy.

Order of Organization (smallest to largest):

Galaxy $\rightarrow$ Local Group $\rightarrow$ Cluster $\rightarrow$ Supercluster $\rightarrow$ Universe

4.3

Cosmology

Redshift
Quasars
Hubble
Dark Matter
The Big Bang
CMB
Universe Models

The Doppler Effect

The Doppler Effect is the apparent change of wavelength which occurs as a celestial body moves away or towards an observer. A celestial object moving away from the Earth appears to have more red lines in its spectrum (red shift) whilst a body moving towards the Earth appears to have more blue in its spectrum (blue shift). The Doppler Effect allows the movement and radial velocity of a celestial body, typically a star, to be measured. This is done by analysing the spectrum of light using a spectrometer and observing how it changes over time. If the wavelengths become shorter and more red shift is observed then the object is moving away from the observer.

Galaxies typically appear to exhibit red shift when observed from the Earth because Earth is near the centre of an expanding universe. Galaxies located within our Local Group show blue shift because we are gravitationally bound, so all the galaxies are moving towards each other. The radial velocity of an object is its velocity along an observer's line of sight; the radial velocity of a star can be calculated by measuring changes in a star's spectrum as it shifts.

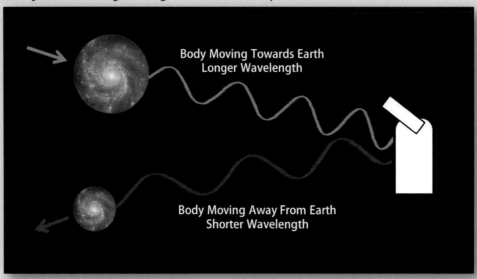

Quasars

The initial discovery of quasars (quasi-stellar objects) occurred during the 1950s when astronomers were able to identify them due to their high red shift and emission of radio waves.

Quasars share a similar appearance to stars, however their behaviour and properties are very different. The actual appearance and properties of quasars are mostly unknown, however, astronomers believe them to be galaxies that emit large quantities of x-rays, ultraviolet radiation and sometimes radio waves.

It is believed that the formation of a quasar occurs as matter falls into a black hole situated at the centre of a galaxy. During this time, jets of matter are propelled out at great velocity which in turn cause their unique appearance.

Through observations, astronomers have been able deduce that quasars show a high red shift, indicating that they are located at great distances from Earth and therefore are likely to be the oldest objects found within our Universe.

Artist's impression of a quasi-stellar object (quasar)

Hubble's Law and Constant

Specification:

- *Describe the mechanics of Hubble's Law and how it explains a relationship between distance and redshift*

- *Explain how the age of the universe is estimated using Hubble's constant*

In 1929 Edwin Hubble combined the recently-discovered concept of the red shift of galaxies as a measure of their velocity away from Earth with his own measurements of distance to formulate Hubble's Law, which states that the velocity of recession of distant galaxies from our own is directly proportional to their distance away from us. Hubble's Law can be expressed using the formula
V=Hd
where:

        V is the recessional velocity of a particular galaxy
        H is the Hubble constant
        d is the distance from the galaxy to the observer

When the redshift and distance from Earth of different distant galaxies are plotted on a graph, the slope of the line giving the best fit is the Hubble Constant. Using the known relationship between distance, velocity and time, Hubble's formula can be rearranged to show that the inverse of Hubble's Constant gives an estimate of the age of the Universe following the Big Bang.

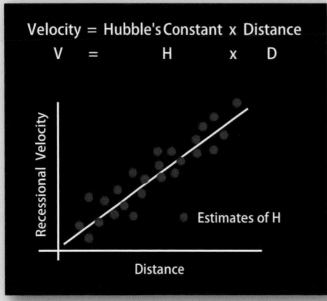

Graph used to determine the value of the Hubble Constant

Dark Matter and Energy

Specification:

- *Describe how astronomers hypothesised the existence of dark matter.*

- *Explain what effect the existence of dark matter is thought to have on the Universe*

Dark Matter

Astronomers are able to calculate the mass of galaxies using Newton's Law of Gravitation. However, it is becoming increasingly apparent that there is a lot more mass in galaxies than we can visually detect. This means that there must be a form of matter which adds mass to galaxies but which cannot yet be detected. Astronomers have proposed that this apparently 'missing' matter exists as some kind of substance which we cannot currently detect using modern technology. This problem has been compared to using your eyes to look at a black object in a dark room, so the matter has been called 'dark matter'. Although dark matter's existence is highly debated, some astronomers believe that up to 95% of the universe's mass could be dark matter.

Dark Energy

Gravity is the force responsible for nearby objects moving towards each other such as the Milky Way and the Andromeda Galaxy, which will collide around 4 billion years from now. However the Universe is known to be expanding, with the clusters of galaxies moving away from each other. The rate of this expansion should have slowed down since the Big Bang, however it is actually speeding up which may be due to an energy which we cannot yet detect. The 'dark energy' hypothesis states that there is a uniformly dispersed dark energy between clusters of galaxies which accelerates the rate of expansion of the Universe.

'What is there in places almost empty of Matter and whence is it that the Sun and Planets gravitate towards one another without dense matter between them?'
Isaac Newton (1832)

Cosmic Microwave Background Radiation

Specification:

- *Describe the nature and hypothesised origin of cosmic microwave background radiation*

- *Explain what effect the existence of cosmic microwave background radiation has on the Universe*

- *Describe recent methods used to observe cosmic microwave background radiation*

Cosmic microwave background radiation (CMBR) was accidentally discovered by the astronomers Penzias and Wilson in 1965 and was initially believed to be man-made, but was afterwards hypothesised to be a remnant of the Big Bang.

After the Big Bang event, the average temperature of the Universe lowered from 3000K to around 3K today. This change also lengthened the wavelengths of electromagnetic waves from 1 millionth of a meter to 1 millimetre which is in the microwave section of the spectrum.

In 1989, NASA sent the Cosmic Background Explorer into space to detect the variation of CMBR in the Universe and created a 'hotspot image' which showed an uneven distribution of CMBR throughout space. This shows that in the early period of the Universe, mass was unevenly distributed

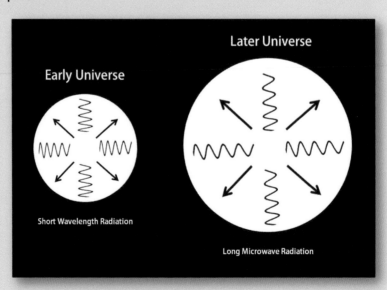

Universe Models

Specification:

- Compare different theories that exist on the lifecycle of the Universe

- Describe the validity of the different hypotheses and evidence that supports the leading theories

A Universe model is a proposal made by astronomers which aims to explain the 'history' of the Universe as well as its present mechanics and future state. There have been a number of these models throughout history and the scientific community still has not entirely agreed on one particular explanation due to a lack of overwhelming evidence in favour of either of the two leading theories.

Early Ideas: Before the discovery of outer space and other galaxies, it was commonly believed that the Milky Way was the entire Universe, with the Earth at the centre of it due to the Church's influence over all aspects of science. However due to the work of Copernicus, Kepler and Galileo, the geocentric model was proved to be wrong, because other galaxies were discovered and the relationship between Earth and other planets as they rotate around the Sun was clarified.

The Steady State Theory: This theory proposed that as the Universe expands, new matter is created so that the overall appearance of the Universe never changes. It went further by also suggesting that the Universe had no beginning and would never have an end.

The Big Bang Theory: This theory proposed that all matter in the Universe was held in one singularity which was at the centre of the present day Universe. A 'big bang' is then thought to have occurred which set off the rapid expansion of the Universe which the astronomer Edwin Hubble concluded is still occurring today.

The most convincing evidence in favour of the Big Bang theory is the existence of cosmic microwave background radiation and the fact that distant galaxies show redshift, proving that the Universe is still expanding, with galactic groups moving further away from each other.

Calculations

Kepler's Laws

Drake's Equation

Declination

Magnitude

The Equation of Time

Kepler's Laws

Specification:

- *Demonstrate Kepler's second law, using a diagram if necessary*

- *Use Kepler's third law in simple calculations*

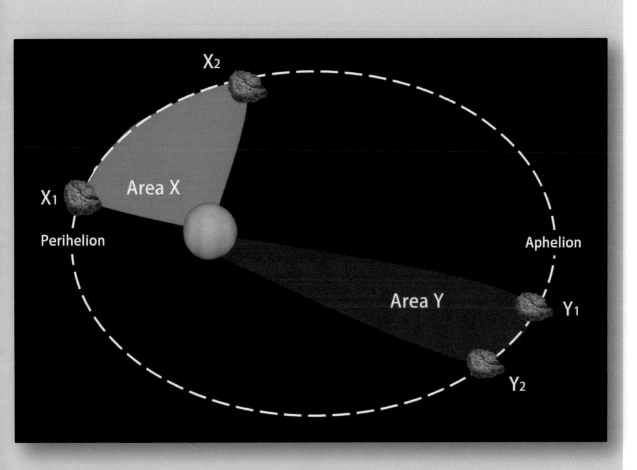

A diagram illustrating Kepler's second law of planetary motion. Kepler's second law states that a planet orbiting the Sun sweeps out an equal area in an equal amount of time

Kepler's third law: The orbital period or planetary distance can be determined as T^2 is directly proportional to r^3, where T is in years, and R is in AU.

Drake Equation

The Drake Equation was proposed by Dr Frank Drake in 1960. The equation was designed as a model for scientists to estimate the probability of life elsewhere in the Universe. Additionally, it further takes into account the possibility of that life being intelligent. The equation is as follows:

$$N = R* \times f_p \times n_e \times f_e \times f_i \times f_c \times L$$

N is the number of civilizations in our galaxy with which communication might be possible;

$R*$ is the average rate of star formation in our galaxy
fp is the fraction of those stars that have planets
n_e is the average number of planets that can potentially support life per star that has planets
f_ℓ is the fraction of the above that actually go on to develop life at some point
f_i is the fraction of the above that actually go on to develop intelligent life
f_c is the fraction of civilizations that develop a technology that releases detectable signs of their existence into space
L is the length of time such civilizations release detectable signals into space.

Calculations

Declination Calculation

The declination of a star is its angle above the celestial equator. Declination is essentially the latitude of a star when measured from the celestial equator. Declination can also be used to establish whether a star is circumpolar.

If it is true that declination > (90 - latitude) then the star is circumpolar. If a star is positively circumpolar it never sets below the horizon.

Magnitude Calculation

Amateur astronomers are only able to observe apparent magnitudes as luminosity decreases with distance, so the formula to calculate absolute magnitude is:

$$M = m + 5 - 5(Log_{10}d)$$

Where M is absolute magnitude, m is apparent magnitude and d is distance in parsecs. Absolute magnitude is the magnitude (brightness) of a star at a distance of 10 parsecs from Earth. The value of absolute magnitude allows comparisons to be made and other variables to be calculated.

The Equation of Time Calculation

The Equation of Time (EoT) is defined as the difference between mean solar time and apparent solar time:
Equation of time = Apparent Solar Time - Mean Solar Time

The EoT changes throughout the year; it can be used to position the observer. Example: A student notes that on the 3rd April the EoT is +4. If a sundial in Greenwich (longitude 0 degrees) reads 12 noon, what is the student's mean solar time (watch time)?
EoT = Apparent Solar Time – Mean Solar Time
Mean Solar Time = Apparent Solar Time - EoT
Mean Solar Time = 12:00 - 00:04
Mean Solar Time = 11:56

Calculations

Recessional Velocity Calculation

By comparing the wavelength of a known part of the electromagnetic spectrum being emitted by a galaxy with the recorded shifted wavelength from the same galaxy, its recessional velocity (the speed that it is moving away from the observer) can be calculated using the following formula:

$$\frac{\lambda - \lambda_o}{\lambda_o} = \frac{\Delta\lambda}{\lambda} = \frac{v}{c}$$

where:

λ = redshifted wavelength

λ_o = unshifted wavelength

$\Delta\lambda$ = change in wavelength

v = recessional velocity

c = speed of light (300,000,000 m/s)

Example: Calculate the recessional velocity of Galaxy X

Hydrogen emits 21.106 cm radio waves which have been shifted by Galaxy X to 21.133 cm by the doppler effect as it moves away from Earth.

$$\Delta\lambda = 21.133 - 21.106 = 0.027 \text{ cm}$$

$$\Delta\lambda/\lambda = v/c = (0.027/21.106) \times 300,000,000 = 380,000 \text{ m/s}$$

Magnification

1. The magnification of a telescope is related to the focal lengths of the objective and the eyepiece by the formula $M = F/f$
where M=Magnification
    F=focal length of objective
    f=focal length of eyepiece
2. The difference in light collected by a telescope is proportional to the square of the power difference of the objective lenses in question.
Example: What is the difference in light collected by a 20cm diameter objective lens compared to a 5cm diameter objective lens?
$20/5 = 4 \rightarrow 4^2 = 16$
so 16 times more light is collected by a 20cm diameter objective lens than a 5cm diameter objective lens.

Example Questions

Topic 1
- What is the approximate circumference of the Earth?
- Define the terms zenith and meridian.
- Which two forms of wavelength pass entirely through the Earth's atmosphere?
- Why are the world's largest telescopes reflector type rather than refractors?
- Explain the key advantages of space based telescopes.
- Why does the rotational period of the Moon differ from its phase period?
- How long is the Sun's solar cycle?
- What is the Equation of Time?

Topic 2
- Define what is meant by the term 'transit'.
- What is an astronomical unit?
- Describe the possible benefits/drawbacks of a manned space mission.
- Describe the difference between a meteoroid, meteor and meteorite.
- How and by whom was Uranus discovered?
- Describe two methods of exoplanet detection.

Topic 3
- What are constellations and how do they differ from an asterism?
- How can long-exposure photography be used to calculate the Earth's rotational period?
- Why do stars appear to move from east to west as they cross the sky?
- An open cluster contains a few young stars. How does this differ from a globular cluster?
- Define the terms absolute and apparent magnitude.
- The Sun's life cycle consists of five stages. What are they?

Topic 4
- What is the diameter of our galaxy, the Milky Way?
- What can be used to easily classify galaxies. What class is the Milky Way?
- Why do galaxies within our Local Group appear to blue shift?
- There are three elements in the radial velocity equation. What are they?

Answers can be found on page 135

Answers

Topic 1

- The approximate circumference of the Earth is 40 000 km.
- Zenith: The point in the sky directly above an observer. Meridian: A circle of set longitude that passes through both poles and the zenith of an observer.
- Radio waves and visible light.
- Large mirrors are easier to manufacture than large lenses.
- No atmosphere to distort the clarity. No light pollution. Weather cannot interfere with observations. Longer periods of darkness.
- The Moon's phase period is 29 days, rather than 27.3 days, due to the movement of the Earth around the Sun.
- 11 years
- The measured difference between apparent solar time and mean solar time.

Topic 2

- A transit occurs when one planet passes in front of another. Commonly used to describe when Venus or Mercury pass in front of the Sun.
- 1 Astronomical Unit (AU) and is around 150000000 km.
- Benefits: Humans are better at problem solving. In financial terms people do not cost as much as developing an intelligent computer system. Drawbacks: Life support has to be constantly supplied to keep the crew alive. If the mission fails, then human life is lost. It takes an extended period time to train an astronaut. Astronauts spending a long time in space suffer medical conditions upon their return.
- A meteoroid is a fragment of dust from comets or asteroids. A meteor is a meteoroid that enters Earth's atmosphere. A meteorite is a meteor that lands on the surface of the Earth.
- Uranus was discovered by William Herschel using a telescope.
- Radial Velocity Detection: Exoplanets exert a gravitational pull on the stars creating a 'wobble'. This wobble can be detected. Astrometry Detection: recording the amount the star moves compared to the background. Transit Method Detection: Measuring regular and uniform dips in the star's luminosity. Optical Detection: Using optical techniques.

Topic 3

- Asterism: a pattern of stars within the night sky. Constellation: a pattern of stars which divides up the night sky.
- Star trails captured using long-exposure photography show the path of a star as it moved across the sky. The arc angle can be measured. The following equation can then be used to determine the rotational period: Rotational period = (360 ÷ θ) × t
- Stars appear to move from east to west due to the Earth's counter-clockwise rotation.
- Globular clusters are a collection of hundreds of thousands of stars that are bound together by gravitational forces. Open clusters contain far fewer stars. Stars within open clusters are spaced much further apart due to weaker gravitational forces.
- Apparent magnitude is the measured luminosity of an object when observed from Earth. Absolute magnitude is the measured luminosity of an object as if it were ten parsecs away from Earth.
- Formation, Protostar, Fusion, Red Giant and White Dwarf.

Topic 4

- 100 000 light years
- Hubble Tuning Fork Diagram. The Milky Way is an Sb galaxy.
- Galaxies within our Local Group blue shift because we are gravitationally bound.
- V represents the radial velocity of a particular galaxy. H the Hubble constant. d denotes the distance to the galaxy.

Index

Printed in Great Britain
by Amazon

78976700R00078